A Practical Introduction to Numerical Methods for Materials Scientists and Engineers

David J. Keffer
Professor of Materials Science & Engineering
University of Tennessee
Knoxville, TN

Copyright © 2015 David J. Keffer
All rights reserved.

First Edition
from the Technical Branch of the Poison Pie Publishing House
http://www.poisonpie.com/publishing

ISBN-13: 978-1517356675

This book is dedicated to my colleagues in the Materials Science & Engineering Department at the University of Tennessee, Knoxville, who provided a good home in which this book could be written.

Preface

We live in a world where our time is limited. As materials scientists and engineers, we fulfill the role of problem-solvers in society. Our success in this endeavor depends upon our ability to solve a particular problem within a given amount of time, which in turns depends upon our familiarity with and access to the standard tools of the trade. A rudimentary knowledge of the tools may allow us to provide a crude solution, whereas a more thorough knowledge of the tools may allow us to generate a more elegant and satisfying solution.

Today, many of the problems posed to materials scientists and engineers involve a computational component. Thus the analysis of a problem can be broken down into problem formulation and problem solution. Problem formulation requires creativity to construct a framework in which the material issue under investigation can be property understood. Problem formulation is also the most crucial step because through the formulation of the model we dictate the physics present in the model. If an important element is omitted in this step, the solution of the model either will not provide any insight into the problem at hand or worse yet provide erroneous guidance. Problem formulation is also the most exciting part of any investigation because our imagination can run wild at this point, leading us to ingenious approaches that not only solve the problem but help us to establish the reputation for creativity and excellence toward which we strive.

However, the formulation of the model must be followed by solution of the model. Today, there are standard tools for solving virtually any variety of mathematical equation, due to the cumulative labors of previous generations of mathematicians and computer scientists. With a working familiarity of these tools, the solution of most models should be simply a matter of methodical routine. While it is true that there is the opportunity to invoke our creativity in the solution of more advanced materials problems requiring specialized numerical techniques, there remains a broad swath of materials problems that can be routinely solved using a relatively small toolbox of standard numerical techniques.

There is the opportunity for great empowerment of the student with this approach. What we hope to avoid is a situation where the rigor in the formulation of the model is sacrificed in order to achieve a mathematically simpler expression, which can be solved with a very limited set of tools. What we hope to achieve is a situation where the formulation proceeds without regard for the ease of numerical solution. Once the model is rigorously formulated, the appropriate numerical solution is then identified. Thus the science guides the numerical techniques, rather than the other way around.

The philosophy espoused in this book is to equip the student with a compact but broadly applicable set of practical problem-solving tools such that the student emerges at the end of the course with the belief, "If I can write the model, I can solve the model."

Summary of the Contents of this Book

This book covers:

- numerical differentiation
- numerical integration
- solution of algebraic equations
    - systems of linear algebraic equations
        - eigenanalysis
        - linear regression
    - non-linear algebraic equations
        - single nonlinear algebraic equations
        - systems of nonlinear algebraic equations
        - optimization
- solution of ordinary differential equations
    - single nonlinear ordinary differential equations
    - systems of nonlinear ordinary differential equations
        - initial value problems
        - boundary value problems

This book does not cover:

- solution of partial differential equations
- solution of integral equations
- signal filtering, including Fourier transforms
- countless other more advanced numerical topics

Table of Contents

Preface iv
Summary of the Contents of this Book v
Table of Contents vi
List of Subroutines ix
Chapter 1. Linear Algebra 1
    1.1. Introduction 1
    1.2. Linearity 1
    1.3. Matrix Notation 3
    1.4. The Determinant and Inverse 5
    1.5. Elementary Row Operations 8
    1.6. Rank and Row Echelon Form 10
    1.7. Existence and Uniqueness of a Solution 14
    1.8. Eigenanalysis 20
    1.9. Summary of Logically Equivalent Statements 29
    1.10. Summary of MATLAB Commands 30
    1.11. Problems 31
Chapter 2. Regression 32
    2.1. Introduction 32
    2.2. Single Variable Linear Regression 32
    2.3. The Variance of the Regression Coefficients 34
    2.4. Multivariate Linear Regression 36
    2.5. Polynomial Regression 38
    2.6 Linearization of Equations 39
    2.7. Confidence Intervals 39
    2.8. Regression Subroutines 41

| 2.9. Problems | 47 |

Chapter 3. Numerical Differentiation 51
- 3.1. Introduction 51
- 3.2. Taylor Series Expansions 51
- 3.3. Finite Difference Formulae 52
- 3.4. Approximations for Partial Derivatives 55
- 3.5. Noise 56
- 3.6. Problems 59

Chapter 4. Solution of a Single Nonlinear Algebraic Equation 60
- 4.1. Introduction 60
- 4.2. Iterative Solutions and Convergence 60
- 4.3. Successive Approximations 61
- 4.4. Bisection Method of Rootfinding 63
- 4.5. Single Variable Newton-Raphson 65
- 4.6. Newton-Raphson with Numerical Derivatives 68
- 4.7. Solution in MATLAB 69
- 4.8. Existence and Uniqueness of Solutions 71
- 4.9. Rootfinding Subroutines 74
- 4.10. Problems 78

Chapter 5. Solution of a System of Nonlinear Algebraic Equations 81
- 5.1. Introduction 81
- 5.2. Multivariate Newton-Raphson Method 81
- 5.3. Multivariate Newton-Raphson Method with Numerical Derivatives 88
- 5.4. Subroutine Codes 89
- 5.5. Problems 91

Chapter 6. Numerical Integration 97
- 6.1. Introduction 97
- 6.2. Trapezoidal Rule 97
- 6.2. Second-Order Simpson's Rule 99
- 6.3. Higher Order Simpson's Rules 102
- 6.5. Quadrature 102
- 6.6. Example 103

6.7. Multidimensional Integrals 106
6.8. Subroutine Codes 109
6.9. Problems 114

Chapter 7. Solution of Ordinary Differential Equations 116
7.1. Introduction 116
7.2. Initial Value Problems 116
7.3. Euler Method 117
7.4. Classical Fourth-Order Runge-Kutta Method 119
7.5. Application to Systems of Ordinary Differential Equations 120
7.6. Higher-Order ODEs 122
7.7. Boundary Value Problems 123
7.8. Subroutine Codes 124
7.9. Problems 131

Chapter 8. Optimization 134
8.1. Introduction 134
8.2. Optimization vs Root-finding in One-Dimension 134
8.4. Other One Dimensional Optimization Techniques 137
8.5. Multivariate Nonlinear Optimization 138
8.6. Optimization vs Root-finding in Multiple Dimensions 139
8.7. Other Multivariate Optimization Techniques 140
8.8. Subroutine Codes 142
8.9. Problems 147

References 149

List of Subroutines

Chapter 1. Linear Algebra
    Summary of Linear Algebra Commands — 30

Chapter 2. Regression
    Single Variable Linear Regression (linreg1) — 41
    Multivariate Linear Regression (linregn) — 42
    Polynomial Regression (polyreg) — 43
    Single Variable Linear Regression with Confidence Intervals (linreg1ci) — 45

Chapter 4. Solution of a Single Nonlinear Algebraic Equation
    Successive Approximation (succapp) — 74
    Bisection (bisect) — 75
    Single Variable Newton-Raphson (newraph1) — 76
    Single Variable Newton-Raphson with Numerical Derivatives (nrnd1) — 77

Chapter 5. Solution of a System of Nonlinear Algebraic Equations
    Multivariate Newton-Raphson with Numerical derivatives (nrndn) — 89

Chapter 6. Numerical Integration
    Trapezoidal Rule (trapezoidal) — 109
    Simpson's Second Order Method (simpson2) — 110
    Simpson's Third Order Method (simpson3) — 111
    Simpson's Fourth Order Method (simpson4) — 111
    Gaussian Quadrature (gaussquad) — 112

Chapter 7. Solution of Ordinary Differential Equations
    Euler Method – 1 Equation IVP (euler1) — 125
    Runge-Kutta Method – 1 Equation IVP (rk41) — 125
    Euler Method – n Equations IVP (eulern) — 126
    Runge-Kutta Method – n Equations IVP (rk4n) — 128
    Runge-Kutta Method – n Equations BVP (rk4n_bvp) — 129

Chapter 8. Optimization
    Bisection Method for optimization – 1 variable (bisect_opt1)   143
    Newton-Raphson Method with numerical derivatives for optimization – 1 variable
        (nrnd_opt1)   144
    Newton-Raphson Method with numerical derivatives for optimization – n variables
        (nrnd_optn)   145

Chapter 1. Linear Algebra

1.1. Introduction

The solution of algebraic equations pervades science and engineering. Thermodynamics is an area rife with examples because of the ubiquitous presence of equilibrium constraints. Thermodynamic constraints are typically algebraic equations. For example, in the two-phase equilibrium of a binary system, there are three algebraic constraints defining the equilibrium state: thermal equilibrium, or equality of temperatures in the two phases, $T^I = T^{II}$; mechanical equilibrium, or equality of pressures in the two phases, $p^I = p^{II}$; and chemical equilibrium, or equality of chemical potentials of component i, in the two phases, $\mu_i^I = \mu_i^{II}$. Under most circumstances, these constraints are algebraic equations because there are no differential or integral operators in the equations.

In the discussion of algebraic equations, it is natural to divide the topic into the solution of linear and nonlinear equations. The mathematical framework for the methodical solution of linear algebraic equations is well-established. There are rigorous techniques for the determination of the existence and uniqueness of solutions. There are established procedures for Eigenanalysis. The discussion of the solution of non-linear algebraic equations is postponed because it is not as straight-forward and requires most of the tools that we develop in our solution of linear algebraic equations.

1.2. Linearity

We have already hinted at the importance of being able to distinguish whether an equation is linear or nonlinear since the solution technique that we adopt is different for linear and nonlinear equations. We begin with a discussion of linear operators. In mathematics, an operator is a symbol or function representing a mathematical operation. Operators that are familiar to undergraduates include exponents, logarithms, differentiation and integration. We can investigate the linearity of each of these operators by applying the following test of linearity

$$L[ax+by] = aL[x] + bL[y] \qquad (1.1)$$

where $L[x]$ is a linear operator, operating on the variable x, y is another variable, and a and b are constants.

We can directly check the four operators listed above for linearity. The differential operator, $L[x(t)] = \frac{d}{dt}[x(t)]$, can be substituted into equation (1.1) to yield

$$\frac{d}{dt}[ax(t) + by(t)] = a\frac{d}{dt}[x(t)] + b\frac{d}{dt}[y(t)] \tag{1.2}$$

The differential operator is indeed linear because we know from differential calculus that a constant can be pulled out of the differential and that the differential of a sum is the sum of the differentials. Similarly, the integral operator, $L[x(t')] = \int_{t_o}^{t}[x(t')]dt'$, can be substituted into equation (1.1) to yield

$$\int_{t_o}^{t}[ax(t') + by(t')]dt' = a\int_{t_o}^{t}[x(t')]dt' + b\int_{t_o}^{t}[y(t')]dt' \tag{1.3}$$

The integral operator is indeed linear because we know from integral calculus that a constant can be pulled out of the integral and that the integral of a sum is the sum of the integrals.

The exponential operator, $L[x] = x^n$, can be substituted into equation (1.1) to yield

$$[ax + by]^n \neq ax^n + by^n \tag{1.4}$$

Equation (1.4) is not generally true. It is true for $n = 1$. However, it is not true for any other integer (positive, negative or zero). We can demonstrate directly that equation (1.4) is not true for $n = 2$ through algebraic manipulation.

$$[ax + by]^2 = a^2x^2 + 2abxy + b^2y^2 \neq ax^2 + by^2 \tag{1.5}$$

Equation (1.4) is also not true for fractional exponents, such as $n = 1/2$.

$$[ax + by]^{1/2} = \sqrt{ax + by} \neq a\sqrt{x} + b\sqrt{y} \tag{1.6}$$

Similarly, the logarithm operator, the inverse operator of the exponential operator is not linear.

$$\log_n(ax + by) \neq a\log_n(x) + b\log_n(y) \tag{1.7}$$

Linear Algebra - 3

Without demonstration, we also state that all trigonometric functions are nonlinear.

In the solution of algebraic equations, the first step is therefore to determine if the equation is linear. An equation is linear if it does not contain any nonlinear operations. If we consider only one equation with one variable, the linear equation has the form,

$$ax = b \tag{1.8.a}$$

which we will choose to rewrite in the more general form of a function as

$$f(x) = ax - b \tag{1.8.b}$$

Examples of nonlinear algebraic equations include

$$f(x) = ax^2 + b \tag{1.9.a}$$

$$f(x) = a\sin(x) + b \tag{1.9.b}$$

$$f(x) = a\exp(x) + x + b \tag{1.9.c}$$

If there is a single nonlinear term in the equation, the entire equation is nonlinear.

In the consideration of systems of equations, if a single equation in the system is nonlinear, then the entire system must be treated as nonlinear.

1.3. Matrix Notation

A system of n algebraic equations containing m unknown variables has the form general form

$$f_i(x) = \sum_{j=1}^{m} a_{i,j} x_j - b_i \qquad \text{for } i = 1 \text{ to } n \tag{1.10}$$

For example, a system with n = 2 algebraic equation and m = 2 variables has the form

$$\begin{aligned} f_1(x_1, x_2) &= a_{1,1} x_1 + a_{1,2} x_2 - b_1 \\ f_2(x_1, x_2) &= a_{2,1} x_1 + a_{2,2} x_2 - b_2 \end{aligned} \tag{1.11}$$

It is conventional to adopt a short-hand notation, known as matrix notation, and express equation (1.10) as

$$\underline{\underline{A}}\,\underline{x} = \underline{b} \tag{1.12}$$

where the matrix of constant coefficients, $\underline{\underline{A}}$, is

$$\underline{\underline{A}} = \begin{bmatrix} a_{1,1} & a_{1,1} & \cdots & a_{1,m} \\ a_{2,1} & a_{2,2} & \cdots & a_{2,m} \\ \vdots & \vdots & \ddots & \vdots \\ a_{n,1} & a_{n,2} & \cdots & a_{n,m} \end{bmatrix} \tag{1.13}$$

The two lines underline in the notation $\underline{\underline{A}}$ indicate that the $\underline{\underline{A}}$ matrix is a two-dimensional matrix. We refer to $\underline{\underline{A}}$ as an *n*x*m* matrix because it contains n rows (equations) and m columns (variables). The vectors $\underline{x}$ and $\underline{b}$ are

$$\underline{x} = \begin{bmatrix} x_1 \\ x_2 \\ \vdots \\ x_m \end{bmatrix} \quad \text{and} \quad \underline{b} = \begin{bmatrix} b_1 \\ b_2 \\ \vdots \\ b_n \end{bmatrix} \tag{1.14}$$

The single underline in the notation $\underline{x}$ and $\underline{b}$ indicates that $\underline{x}$ and $\underline{b}$ are vectors, or one-dimensional matrices. We refer to $\underline{x}$ and $\underline{b}$ as "column vectors" of size *m*x1 and *n*x1 respectively.

    The crucial thing to remember about matrix notation is that it is a convention to simplify the notation. It does not introduce any new mathematical rules. It does not add or change the rules of algebra. The solution to the set of equation can be obtained following the familiar rules of algebra, although such manipulations become cumbersome when the number of equations is large. Therefore we will shortly adopt a set of notations for matrix operations, which consist of a sequence of algebraic rules.

    In equation (1.12), we present the first matrix operation, matrix multiplication, $\underline{\underline{A}}\,\underline{x}$. Matrices can only be multiplied if the inner indices match. In this case $\underline{\underline{A}}$ is of size *n*x*m* and $\underline{x}$ is of size *m*x1. Since the last index of $\underline{\underline{A}}$ matches the first index of $\underline{x}$, they can be multiplied. If the indices do not match, there is not matrix multiplication. For example, the matrix multiplication $\underline{x}\,\underline{\underline{A}}$ cannot be performed because $\underline{x}$ is of size *m*x1 and $\underline{\underline{A}}$ is of size *n*x*m* and the inner indices, 1 and *n* are not the same.

Linear Algebra - 5

The matrix resulting from a valid matrix multiplication is of a size defined by the two outer indices of the factor matrices. Thus $\underline{\underline{A}}\underline{x}$ yields a matrix of size $n \times 1$, a column vector of length n. The ith element of an $n \times m$ matrix, $\underline{\underline{A}}$, and an $m \times 1$ column vector, $\underline{x}$, is defined as

$$\sum_{j=1}^{m} a_{i,j} x_j \qquad \text{for } i = 1 \text{ to } n \tag{1.15}$$

Similarly, the i,j^{th} element of an $n \times m$ matrix, $\underline{\underline{A}}$, and an $m \times p$ column vector, $\underline{\underline{B}}$, is defined as

$$\sum_{k=1}^{m} a_{i,k} b_{k,j} \qquad \text{for } i = 1 \text{ to } n \text{ and for } j = 1 \text{ to } p \tag{1.16}$$

1.4. The Determinant and Inverse

We now consider the solution of equation (1.10) or alternatively equation (1.12). When n=1 equation and m=1 variable, we have

$$a_{1,1} x_1 = b_1 \tag{1.15}$$

This of course has the general solution, $x_1 = b_1 / a_{1,1}$. This simple problem illustrates the issue of existence of a solution. The solution only exists if $a_{1,1} \neq 0$.

We can next consider a set of linear algebraic equations with n=2 equations and m=2 variables, we have

$$\begin{aligned} a_{1,1} x_1 + a_{1,2} x_2 &= b_1 \\ a_{2,1} x_1 + a_{2,2} x_2 &= b_2 \end{aligned} \tag{1.16}$$

Through a series of algebraic manipulations, we can arrive at the solution

$$x_1 = \frac{a_{22} b_1 - a_{12} b_2}{a_{11} a_{22} - a_{21} a_{12}} \qquad \text{and} \qquad x_2 = \frac{-a_{21} b_1 + a_{11} b_2}{a_{11} a_{22} - a_{21} a_{12}} \tag{1.17}$$

Note that both x_1 and x_2 have the same denominator. This denominator is given a special name, the determinant, $\det(\underline{\underline{A}})$.

$$\det(\underline{\underline{A}}_{2 \times 2}) = a_{11} a_{22} - a_{21} a_{12} \tag{1.18}$$

It is clear that a solution only exists if $\det(\underline{\underline{A}}) \neq 0$. This is exactly parallel to the single equation case given above, where the determinant of the one equation case is simply $\det(\underline{\underline{A}}_{1x1}) = a_{11}$.

We can next consider a set of linear algebraic equations with n=3 equations and m=3 variables, we have

$$\begin{aligned} a_{1,1}x_1 + a_{1,2}x_2 + a_{1,3}x_3 &= b_1 \\ a_{2,1}x_1 + a_{2,2}x_2 + a_{2,3}x_3 &= b_2 \\ a_{3,1}x_1 + a_{3,2}x_2 + a_{3,3}x_3 &= b_3 \end{aligned} \quad (1.19)$$

Through a series of algebraic manipulations, we can arrive at the solution

$$\begin{aligned} x_1 &= \frac{(a_{22}a_{33} - a_{32}a_{23})b_1 + (-a_{12}a_{33} + a_{32}a_{13})b_2 + (a_{12}a_{23} - a_{22}a_{13})b_3}{\det(\underline{\underline{A}}_3)} \\ x_2 &= \frac{(-a_{21}a_{33} + a_{31}a_{23})b_1 + (a_{11}a_{33} - a_{31}a_{13})b_2 + (-a_{11}a_{23} + a_{21}a_{13})b_3}{\det(\underline{\underline{A}}_3)} \\ x_3 &= \frac{(a_{21}a_{32} - a_{31}a_{22})b_1 + (-a_{11}a_{32} + a_{31}a_{12})b_2 + (a_{11}a_{22} - a_{21}a_{12})b_3}{\det(\underline{\underline{A}}_3)} \end{aligned} \quad (1.20)$$

where

$$\det(\underline{\underline{A}}_{3x3}) = a_{11}(a_{22}a_{33} - a_{32}a_{23}) + a_{12}(a_{23}a_{31} - a_{33}a_{21}) + a_{13}(a_{21}a_{32} - a_{31}a_{22}) \quad (1.21)$$

Note that both x_1, x_2 and x_3 have the same denominator. It is clear that a solution only exists if $\det(\underline{\underline{A}}) \neq 0$.

There is no theoretical reason that we could not continue to solve systems of *n* linear algebraic equations with *n* variables for arbitrary *n*. However, practically speaking it becomes very time consuming. Note in all these cases that the determinant is strictly a function of $\underline{\underline{A}}$. The determinant is not a function of $\underline{x}$ and $\underline{b}$. At this point, we simply extrapolate the mathematical observation that **a unique solution to $\underline{\underline{A}}\underline{x} = \underline{b}$ exists only if the determinant of the matrix $\underline{\underline{A}}$ exists.**

It turns out that the solution to the 2x2 problem given in equation (1.17) and the solution to the 3x3 problem given in equation (1.20) and the solution for the general *n*x*n* problem can be expressed in matrix notation as

Through a series of algebraic manipulations, we can arrive at the solution

$$\underline{x} = \underline{\underline{A}}^{-1}\underline{b} \quad (1.22)$$

where $\underline{\underline{A}}^{-1}$ is called the inverse matrix of $\underline{\underline{A}}$. In addition to providing the solution to $\underline{\underline{A}}\underline{x} = \underline{b}$ as given in equation (1.22), the inverse all has the additional property,

$$\underline{\underline{A}}^{-1}\underline{\underline{A}} = \underline{\underline{A}}\underline{\underline{A}}^{-1} = \underline{\underline{I}} \tag{1.23}$$

where $\underline{\underline{I}}$ is the identity matrix, defined as

$$\underline{\underline{I}} \equiv \begin{bmatrix} 1 & 0 & \cdots & 0 \\ 0 & 1 & \cdots & 0 \\ \vdots & \vdots & \ddots & \vdots \\ 0 & 0 & \cdots & 1 \end{bmatrix} \tag{1.24}$$

One can also derive equation (1.22) as follows

$$\underline{\underline{A}}\underline{x} = \underline{b}$$
$$\underline{\underline{A}}^{-1}\underline{\underline{A}}\underline{x} = \underline{\underline{A}}^{-1}\underline{b}$$
$$\underline{\underline{I}}\underline{x} = \underline{\underline{A}}^{-1}\underline{b}$$
$$\underline{x} = \underline{\underline{A}}^{-1}\underline{b}$$

We can observe directly from the examples given above that for the small 1x1, 2x2 and 3x3 systems,

$$\underline{\underline{A}}^{-1}_{1x1} = \frac{1}{\det(\underline{\underline{A}}_{1x1})} = \frac{1}{a_{11}} \tag{1.25.a}$$

$$\underline{\underline{A}}^{-1}_{2x2} = \frac{1}{\det(\underline{\underline{A}}_{2x2})} \begin{bmatrix} a_{22} & -a_{12} \\ -a_{21} & a_{11} \end{bmatrix} \tag{1.25.b}$$

$$\underline{\underline{A}}^{-1}_{3x3} = \frac{1}{\det(\underline{\underline{A}}_{3x3})} \begin{bmatrix} a_{22}a_{33} - a_{32}a_{23} & -a_{12}a_{33} + a_{32}a_{13} & a_{12}a_{23} - a_{22}a_{13} \\ -a_{21}a_{33} + a_{31}a_{23} & a_{11}a_{33} - a_{31}a_{13} & -a_{11}a_{23} + a_{21}a_{13} \\ a_{21}a_{32} - a_{31}a_{22} & -a_{11}a_{32} + a_{31}a_{12} & a_{11}a_{22} - a_{21}a_{12} \end{bmatrix} \tag{1.25.c}$$

Clearly from equation (1.25), the inverse does not exist if the determinant is zero. If the inverse of a matrix $\underline{\underline{A}}$ exists, $\underline{\underline{A}}$ is said to be non-singular. If the inverse of the matrix $\underline{\underline{A}}$ does not exist, $\underline{\underline{A}}$ is said to be singular.

1.5. Elementary Row Operations

There exists a methodical procedure for generating inverses analytically. With the ubiquitous presence of computers, it is unlikely that any student will ever have any need to perform such a procedure. It is not even perfectly clear that it is essential to include such a procedure in a modern textbook. Nevertheless, since students may be called upon to generate inverses of small systems in an examination in which computers are not available, we present the procedure here.

The procedure uses three elementary row operations. The first elementary row operation is the multiplication of a row by a constant.

$$row\,1 = c \cdot row\,1 \tag{1.26}$$

The second elementary row operation is switching the order of rows. Clearly, the order that the equations are written should not influence the validity of the equations.

$$row\,2 \leftrightarrow row\,1 \tag{1.27}$$

The third elementary row operation is the replacement of an equation by the linear combination of that equation with other equations. In other words, either equation in (1.28) can be replaced by

$$row\,2 = a \cdot row\,1 + b \cdot row\,2 \tag{1.28}$$

and the resulting system of equations will still yield the same result.

These three elementary row operations provide the necessary tools to (i) determine the existence and unique of solutions, (ii) determine the inverse of $\underline{\underline{A}}$ if it exists and (iii) provide the solution to $\underline{\underline{A}}\underline{x} = \underline{b}$.

For a 2x2 matrix, the procedure for finding the inverse is given below. First, we create an augmented $\underline{\underline{A}}|\underline{\underline{I}}$ matrix. If $\underline{\underline{A}}$ is an nxn matrix and $\underline{\underline{I}}$ is an nxn identity matrix, then $\underline{\underline{A}}|\underline{\underline{I}}$ is an nx$(2n)$ matrix. defined as

$$\underline{\underline{A}}|\underline{\underline{I}} = \begin{bmatrix} a_{1,1} & \cdots & a_{1,n} & 1 & 0 & 0 \\ \vdots & \ddots & \vdots & 0 & \ddots & 0 \\ a_{n,1} & \cdots & a_{n,n} & 0 & 0 & 1 \end{bmatrix} \tag{1.29}$$

In general, one performs elementary row operations that convert the $\underline{\underline{A}}$ side of the augmented matrix to $\underline{\underline{I}}$. At the same time, one performs the same elementary row operations to the $\underline{\underline{I}}$ side of the augmented matrix, which converts it to the inverse of $\underline{\underline{A}}$.

We can illustrate the process for a 2x2 matrix.

Linear Algebra - 9

$$A|I = \begin{bmatrix} a_{11} & a_{12} & | & 1 & 0 \\ a_{21} & a_{22} & | & 0 & 1 \end{bmatrix}$$

(1) Put a one in the diagonal element of ROW 1.

$$ROW1 = \frac{ROW1}{a_{11}}$$

$$\begin{bmatrix} 1 & a_{12}/a_{11} & | & 1/a_{11} & 0 \\ a_{21} & a_{22} & | & 0 & 1 \end{bmatrix}$$

(2) Put zeroes in all the entries of COLUMN 1 except ROW 1.

$$ROW2 = ROW2 - a_{21} ROW1$$

$$\begin{bmatrix} 1 & a_{12}/a_{11} & | & 1/a_{11} & 0 \\ 0 & a_{22} - a_{21}a_{12}/a_{11} & | & -a_{21}/a_{11} & 1 \end{bmatrix}$$

(3) Put a one in the diagonal element of ROW 2.

$$ROW2 = \frac{ROW2}{a_{22} - a_{21}a_{12}/a_{11}}$$

$$\begin{bmatrix} 1 & a_{12}/a_{11} & | & 1/a_{11} & 0 \\ 0 & 1 & | & \dfrac{-a_{21}/a_{11}}{a_{22} - a_{21}a_{12}/a_{11}} & \dfrac{1}{a_{22} - a_{21}a_{12}/a_{11}} \end{bmatrix}$$

(4) Put zeroes in all the entries of COLUMN 2 except ROW 2.

$$ROW1 = ROW1 - \frac{a_{12}}{a_{11}} ROW2$$

$$\begin{bmatrix} 1 & 0 & \dfrac{1}{a_{11}} - \dfrac{a_{12}}{a_{11}}\left(\dfrac{-a_{21}/a_{11}}{a_{22}-a_{21}a_{12}/a_{11}}\right) & -\dfrac{a_{12}}{a_{11}}\left(\dfrac{1}{a_{22}-a_{21}a_{12}/a_{11}}\right) \\ 0 & 1 & \dfrac{-a_{21}/a_{11}}{a_{22}-a_{21}a_{12}/a_{11}} & \dfrac{1}{a_{22}-a_{21}a_{12}/a_{11}} \end{bmatrix}$$

which can be simplified as:

$$\begin{bmatrix} 1 & 0 & \dfrac{a_{22}}{a_{11}a_{22}-a_{21}a_{12}} & \dfrac{-a_{12}}{a_{11}a_{22}-a_{21}a_{12}} \\ 0 & 1 & \dfrac{-a_{21}}{a_{11}a_{22}-a_{21}a_{12}} & \dfrac{a_{11}}{a_{11}a_{22}-a_{21}a_{12}} \end{bmatrix}$$

Here we have converted the matrix on the left hand side to the identity matrix. The matrix on the right hand side is now the inverse as can be seen through comparison of equation (1.25.b).

We can learn several things about the inverse from this demonstration. The most important thing is that if the determinant is zero, the inverse does not exist (because we divide by the determinant to obtain the inverse.)

> Never calculate an inverse until you have first shown that the determinant is not zero.

1.6. Rank and Row Echelon Form

To determine the existence and uniqueness of the solution to $\underline{\underline{A}}\,\underline{x} = \underline{b}$, we must create an augmented $\underline{\underline{A|b}}$ matrix. If $\underline{\underline{A}}$ is an $n\mathrm{x}n$ matrix and $\underline{b}$ is an $n\mathrm{x}1$ column vector, then $\underline{\underline{A|b}}$ is an $n\mathrm{x}(n+1)$ matrix. defined as

$$\underline{\underline{A|b}} = \begin{bmatrix} a_{1,1} & \cdots & a_{n,1} & b_1 \\ \vdots & \ddots & \vdots & \vdots \\ a_{n,1} & \cdots & a_{n,n} & b_n \end{bmatrix} \qquad (1.30)$$

Linear Algebra - 11

In order to determine the existence and uniqueness of a solution to $\underline{\underline{A}}\underline{x} = \underline{b}$, we need to put the matrix $\underline{\underline{A}}$ and the augmented matrix $\underline{\underline{A}}|\underline{b}$ into **row echelon form** (ref) (or **reduced row echelon form** (rref)) using a sequence of elementary row operations.

Row echelon form is also called upper triangular form, in which all elements below the diagonal are zero. For an arbitrary 2x2 matrix $\underline{\underline{A}}$, we have

$$\underline{\underline{A}}_{2x2} = \begin{bmatrix} a_{11} & a_{12} \\ a_{21} & a_{22} \end{bmatrix}$$

We can put this matrix into row echelon form with one elementary row operation, namely

$$ROW\ 2 = ROW\ 2 - \frac{a_{21}}{a_{11}} ROW\ 1$$

which yields

$$ref\left(\underline{\underline{A}}_{2x2}\right) = \begin{bmatrix} a_{11} & a_{12} \\ 0 & a_{22} - \frac{a_{21}}{a_{11}} a_{12} \end{bmatrix}$$

Equation (1.32) is the row echelon form of $\underline{\underline{A}}$. Reduced row echelon form simply requires dividing each row of the row echelon form by the diagonal element of that row,

$$ROW\ 1 = \frac{1}{a_{11}} ROW\ 1$$

$$ROW\ 2 = \frac{1}{a_{22} - \frac{a_{21}}{a_{11}} a_{12}} ROW\ 2$$

which yields

$$rref\left(\underline{\underline{A}}_{2x2}\right) = \begin{bmatrix} 1 & a_{12}/a_{11} \\ 0 & 1 \end{bmatrix}$$

Equation (1.33) is the reduced row echelon form of $\underline{\underline{A}}$ because it is in row echelon form and it has ones in the diagonal elements.

Similarly, the augmented $\underline{\underline{A|b}}$ matrix can be put into row echelon form or reduced row echelon form using precisely the same set of elementary row operations. For a 2x2 example, we have for the row echelon form

$$\underline{\underline{A|b}} = \begin{bmatrix} a_{1,1} & a_{2,1} & \Big| & b_1 \\ a_{2,1} & a_{2,2} & \Big| & b_2 \end{bmatrix}$$

$$ROW\ 2 = ROW\ 2 - \frac{a_{21}}{a_{11}} ROW\ 1$$

$$ref\left(\underline{\underline{A|b}}\right) = \begin{bmatrix} a_{1,1} & a_{2,1} & \Big| & b_1 \\ 0 & a_{2,2} - \frac{a_{21}}{a_{11}} a_{12} & \Big| & b_2 - \frac{a_{21}}{a_{11}} b_1 \end{bmatrix}$$

For a 2x2 example, we have for the reduced row echelon form

$$ROW\ 1 = \frac{1}{a_{11}} ROW\ 1$$

$$ROW\ 2 = \frac{1}{a_{22} - \frac{a_{21}}{a_{11}} a_{12}} ROW\ 2$$

$$rref\left(\underline{\underline{A|b}}\right) = \begin{bmatrix} 1 & a_{2,1}/a_{1,1} & \Big| & b_1/a_{1,1} \\ 0 & 1 & \Big| & \frac{b_2 - \frac{a_{21}}{a_{11}} b_1}{a_{2,2} - \frac{a_{21}}{a_{11}} a_{12}} \end{bmatrix} = \begin{bmatrix} 1 & a_{2,1}/a_{1,1} & \Big| & b_1/a_{1,1} \\ 0 & 1 & \Big| & \frac{a_{11} b_2 - a_{21} b_1}{\det\left(\underline{\underline{A}}\right)} \end{bmatrix}$$

In order to evaluate the existence and uniqueness of a solution, we also require the **rank** of a matrix. The rank of a matrix $\underline{\underline{A}}$ is the number of non-zero rows in a matrix when it is put in row echelon form. The rank of a matrix in row echelon form is the same as the rank of a matrix in reduced row echelon form.

Consider the following upper triangular matrices.

$$\underline{\underline{U}} = \begin{bmatrix} u_{11} & u_{12} & u_{13} \\ 0 & u_{22} & u_{23} \\ 0 & 0 & u_{33} \end{bmatrix} \quad (1.31)$$

The rank of this matrix is 3. The determinant of this matrix is non-zero.

If the determinant of an nxn matrix is zero, then the $rank(\underline{\underline{A}}_n)$ is less than n.

$$\underline{\underline{U}} = \begin{bmatrix} u_{11} & u_{12} & u_{13} \\ 0 & u_{22} & u_{23} \\ 0 & 0 & 0 \end{bmatrix} \quad (1.32)$$

Non-square matrices can also be put in row echelon form. Consider the augmented nx(n+1) matrix of the form:

$$\underline{\underline{U}} = \begin{bmatrix} u_{11} & u_{12} & u_{13} & | & v_1 \\ 0 & u_{22} & u_{23} & | & v_2 \\ 0 & 0 & u_{33} & | & v_3 \end{bmatrix} \quad (1.33)$$

The rank of this matrix is still defined as the number of non-zero rows in the row echelon form of the matrix. The rank of the matrix shown above is 3. For augmented matrices, the non-zero element can appear on either matrix. The rank of the following matrix is still three.

$$\underline{\underline{U}} = \begin{bmatrix} u_{11} & u_{12} & u_{13} & | & v_1 \\ 0 & u_{22} & u_{23} & | & v_2 \\ 0 & 0 & 0 & | & v_3 \end{bmatrix} \quad (1.34)$$

In an augmented matrix, both sides of a row must be zero for the row to be considered zero. The rank of the following matrix is two.

$$\underline{\underline{U}} = \begin{bmatrix} u_{11} & u_{12} & u_{13} & | & v_1 \\ 0 & u_{22} & u_{23} & | & v_2 \\ 0 & 0 & 0 & | & 0 \end{bmatrix} \quad (1.35)$$

The rank provides the number of independent equations in the system. For example, consider the 3x3 example given below. The third equation is a linear combination of the first two equations. This matrix can be put in row echelon form using the following elementary row operations,

$$\underline{\underline{A}} = \begin{bmatrix} a_{11} & a_{12} & a_{13} \\ a_{21} & a_{22} & a_{23} \\ ca_{11} + ka_{21} & ca_{12} + ka_{22} & ca_{13} + ka_{23} \end{bmatrix} \quad (1.36)$$

This matrix can be put in row echelon form using the following elementary row operations,

$$ROW\ 3 = ROW\ 3 - cROW\ 1 - kROW\ 2$$

$$ROW\ 2 = ROW\ 2 - \frac{a_{21}}{a_{11}} ROW\ 1$$

which yields

$$(\underline{\underline{A}}) = \begin{bmatrix} a_{11} & a_{12} & a_{13} \\ 0 & a_{22} - \frac{a_{21}}{a_{11}} a_{12} & a_{23} - \frac{a_{21}}{a_{11}} a_{13} \\ 0 & 0 & 0 \end{bmatrix} \quad (1.37)$$

1.7. Existence and Uniqueness of a Solution

The existence and uniqueness of a solution to $\underline{\underline{A}}\underline{x} = \underline{b}$ can be determined with either the row echelon form or the reduced row echelon form of $\underline{\underline{A}}$ and $\underline{\underline{A}}|\underline{b}$ as follows. In dealing with linear equations, we only have three choices for the number of solutions. We either have 0, 1, or an infinite number of solutions.

No Solutions:
$$rank(\underline{\underline{A}}) < n \text{ and } rank(\underline{\underline{A}}) < rank(\underline{\underline{A}}|\underline{b})$$

One Solution:
$$rank(\underline{\underline{A}}) = rank(\underline{\underline{A}}|\underline{b}) = n$$

Infinite Solutions:
$$rank(\underline{\underline{A}}) = rank(\underline{\underline{A}}|\underline{b}) < n$$

Linear Algebra - 15

When $rank(\underline{\underline{A}}|\underline{b}) > rank(\underline{\underline{A}})$, your system is over-specified. At least one of the equations is linearly dependent in the a matrix but is assigned to an inconsistent value of in the $\underline{b}$ vector. There are no solutions to your problem. When $rank(\underline{\underline{A}}|\underline{b}) = rank(\underline{\underline{A}}) = n$, you have a properly specified system with n equations and n unknown variables and you have one, unique solution. When $rank(\underline{\underline{A}}|\underline{b}) = rank(\underline{\underline{A}}) < n$, then you have less equations than unknowns. You can pick $n - rank(\underline{\underline{A}})$ unknowns arbitrarily then solve for the rest. Therefore you have an infinite number of solutions. We will work one example of each case below.

Example 1.1. One Solution to $\underline{\underline{A}}\underline{x} = \underline{b}$

Let's find
    (a) the determinant of $\underline{\underline{A}}$
    (b) the inverse of $\underline{\underline{A}}$
    (c) the solution of $\underline{\underline{A}}\underline{x} = \underline{b}_1$
    (d) the solution of $\underline{\underline{A}}\underline{x} = \underline{b}_2$

where

$$\underline{\underline{A}} = \begin{bmatrix} 2 & 1 & 3 \\ 1 & 2 & 1 \\ 1 & 1 & 1 \end{bmatrix} \quad \underline{b}_1 = \begin{bmatrix} 1 \\ 1 \\ 1 \end{bmatrix} \quad \underline{b}_2 = \begin{bmatrix} 2 \\ 0 \\ 2 \end{bmatrix}$$

Solution:
(a) The determinant of $\underline{\underline{A}}$ is (by equation 1.21) $\det(\underline{\underline{A}}) = -1$.
(b) Because the determinant is non-zero, we know there will be an inverse. Let's find it.
STEP ONE. Write down the initial matrix augmented by the identity matrix.

$$\underline{\underline{A}}|I = \begin{bmatrix} 2 & 1 & 3 & 1 & 0 & 0 \\ 1 & 2 & 1 & 0 & 1 & 0 \\ 1 & 1 & 1 & 0 & 0 & 1 \end{bmatrix}$$

STEP TWO. Using elementary row operations, convert $\underline{\underline{A}}$ into an identity matrix.

(1) Put a one in the diagonal element of ROW 1. $ROW1 = \dfrac{ROW1}{a_{11}} = \dfrac{ROW1}{2}$

$$\begin{bmatrix} 1 & 1/2 & 3/2 & | & 1/2 & 0 & 0 \\ 1 & 2 & 1 & | & 0 & 1 & 0 \\ 1 & 1 & 1 & | & 0 & 0 & 1 \end{bmatrix}$$

(2) Put zeroes in all the entries of COLUMN 1 except ROW 1.

$$ROW\,2 = ROW\,2 - a_{21}ROW\,1 = ROW\,2 - ROW\,1$$
$$ROW\,3 = ROW\,3 - a_{31}ROW\,1 = ROW\,3 - ROW\,1$$

$$\begin{bmatrix} 1 & 1/2 & 3/2 & | & 1/2 & 0 & 0 \\ 0 & 3/2 & -1/2 & | & -1/2 & 1 & 0 \\ 0 & 1/2 & -1/2 & | & -1/2 & 0 & 1 \end{bmatrix}$$

(3) Put a 1 in the diagonal element of ROW 2. $ROW\,2 = \dfrac{ROW\,2}{a_{22}} = \dfrac{ROW\,2}{3/2}$

$$\begin{bmatrix} 1 & 1/2 & 3/2 & | & 1/2 & 0 & 0 \\ 0 & 1 & -1/3 & | & -1/3 & 2/3 & 0 \\ 0 & 1/2 & -1/2 & | & -1/2 & 0 & 1 \end{bmatrix}$$

(4) Put zeroes in all the entries of COLUMN 2 except ROW 2.

$$ROW\,1 = ROW\,1 - a_{12}ROW\,2 = ROW\,1 - 1/2 * ROW\,2$$
$$ROW\,3 = ROW\,3 - a_{32}ROW\,2 = ROW\,3 - 1/2 * ROW\,2$$

$$\begin{bmatrix} 1 & 0 & 5/3 & | & 2/3 & -1/3 & 0 \\ 0 & 1 & -1/3 & | & -1/3 & 2/3 & 0 \\ 0 & 0 & -1/3 & | & -1/3 & -1/3 & 1 \end{bmatrix}$$

(5) Put a 1 in the diagonal element of ROW 3. $ROW\,3 = \dfrac{ROW\,3}{a_{33}} = \dfrac{ROW\,3}{-1/3}$

$$\begin{bmatrix} 1 & 0 & 5/3 & | & 2/3 & -1/3 & 0 \\ 0 & 1 & -1/3 & | & -1/3 & 2/3 & 0 \\ 0 & 0 & 1 & | & 1 & 1 & -3 \end{bmatrix}$$

Linear Algebra - 17

(6) Put zeroes in all the entries of COLUMN 3 except ROW 3.

$$ROW1 = ROW1 - a_{13}ROW3 = ROW1 - 5/3 * ROW3$$
$$ROW2 = ROW2 - a_{23}ROW3 = ROW2 + 1/3 * ROW3$$

$$\left[\begin{array}{ccc|ccc} 1 & 0 & 0 & -1 & -2 & 5 \\ 0 & 1 & 0 & 0 & 1 & -1 \\ 0 & 0 & 1 & 1 & 1 & -3 \end{array}\right]$$

We have the inverse.

$$\underline{\underline{A}}^{-1} = \begin{bmatrix} -1 & -2 & 5 \\ 0 & 1 & -1 \\ 1 & 1 & -3 \end{bmatrix}$$

(c) The solution to $\underline{\underline{A}}\underline{x} = \underline{b}_1$ is $\underline{x} = \underline{\underline{A}}^{-1}\underline{b}_1$.

$$\underline{x} = \begin{bmatrix} -1 & -2 & 5 \\ 0 & 1 & -1 \\ 1 & 1 & -3 \end{bmatrix} \begin{bmatrix} 1 \\ 1 \\ 1 \end{bmatrix} = \begin{bmatrix} 2 \\ 0 \\ -1 \end{bmatrix}$$

(d) The solution to $\underline{\underline{A}}\underline{x} = \underline{b}_2$ is $\underline{x} = \underline{\underline{A}}^{-1}\underline{b}_2$.

$$\underline{x} = \begin{bmatrix} -1 & -2 & 5 \\ 0 & 1 & -1 \\ 1 & 1 & -3 \end{bmatrix} \begin{bmatrix} 2 \\ 0 \\ 2 \end{bmatrix} = \begin{bmatrix} 8 \\ -2 \\ -4 \end{bmatrix}$$

We see that we only need to calculate the inverse once to solve both $\underline{\underline{A}}\underline{x} = \underline{b}_1$ and $\underline{\underline{A}}\underline{x} = \underline{b}_2$. That's nice because finding the inverse is a lot harder than solving the equation once the inverse is known.

Example 1.2. No Solutions to $\underline{\underline{A}}\underline{x} = \underline{b}$

$$\underline{\underline{A}} = \begin{bmatrix} 2 & 1 & 3 \\ 1 & 2 & 1 \\ 3 & 3 & 4 \end{bmatrix} \quad \underline{b}_1 = \begin{bmatrix} 1 \\ 1 \\ 1 \end{bmatrix}$$

In this case, when we compute the determinant, we find that $\det(\underline{\underline{A}}) = 0$. The determinant is zero. No inverse exists. To determine if we have no solution or infinite solutions find the ranks of $\underline{\underline{A}}$ and $\underline{\underline{A|b}}$. In row echelon form, A becomes:

$$ref(\underline{\underline{A}}) = \begin{bmatrix} 2 & 1 & 3 \\ 0 & -3 & 1 \\ 0 & 0 & 0 \end{bmatrix}$$

By inspection of the row echelon form, $rank(\underline{\underline{A}}) = 2$. In row echelon form, $\underline{\underline{A|b}}$ becomes

$$ref(\underline{\underline{A|b}}) = \begin{bmatrix} 2 & 1 & 3 & 1 \\ 0 & -3 & 1 & -1 \\ 0 & 0 & 0 & -2 \end{bmatrix}$$

By inspection of the row echelon form, $rank(\underline{\underline{A|b}}) = 3$. Since $rank(\underline{\underline{A|b}}) > rank(\underline{\underline{A}})$, there are no solutions to $\underline{\underline{A}}\underline{x} = \underline{b}$.

Example 1.3. Infinite Solutions to $\underline{\underline{A}}\underline{x} = \underline{b}$

Consider the same matrix, $\underline{\underline{A}}$, as was used in the previous example 1.2. The determinant is zero and the rank is 2. Now consider a different b vector.

$$\underline{b}_2 = \begin{bmatrix} 1 \\ 1 \\ 2 \end{bmatrix}$$

In row echelon form, $\underline{\underline{A|b}}$ becomes:

$$ref\left(\underline{\underline{A}}|\underline{b}\right) = \begin{bmatrix} 2 & 1 & 3 & 1 \\ 0 & -3 & 1 & -1 \\ 0 & 0 & 0 & 0 \end{bmatrix}$$

By inspection of the row echelon form, $rank(A|b) = 2$. Since $rank(\underline{\underline{A}}) = rank(\underline{\underline{A}}|\underline{b}) = 2 < n = 3$, there are infinite solutions.

We can find one example of the infinite solutions by following a standard procedure. First, we arbitrarily select $n - rank(\underline{\underline{A}})$ variables. In this case we can select one variable. Let's make $x_3 = 0$. Then substitute that value into the row echelon form of $\underline{\underline{A}}|\underline{b}$ and solve the resulting system of $rank(\underline{\underline{A}})$ equations.

$$ref\left(\underline{\underline{A}}|\underline{b}\right) = \begin{bmatrix} 2 & 1 & 3 & 1 \\ 0 & -3 & 1 & -1 \\ 0 & 0 & 0 & 0 \end{bmatrix}$$

When $x_3 = 0$

$$ref\left(\underline{\underline{A}}|\underline{b}\right) = \begin{bmatrix} 2 & 1 & 0 & 1 \\ 0 & -3 & 0 & -1 \\ 0 & 0 & 0 & 0 \end{bmatrix}$$

Now solve a new $\underline{\underline{A}}\underline{x} = \underline{b}$ problem where $\underline{\underline{A}}$ and $\underline{b}$ come from the non-zero parts of $ref\left(\underline{\underline{A}}|\underline{b}\right)$.

$$\begin{bmatrix} 2 & 1 \\ 0 & -3 \end{bmatrix} \begin{bmatrix} x_1 \\ x_2 \end{bmatrix} = \begin{bmatrix} 1 \\ -1 \end{bmatrix}$$

This problem will always have an inverse. The solution is given by

$$\begin{bmatrix} x_1 \\ x_2 \end{bmatrix} = \begin{bmatrix} 1/3 \\ 1/3 \end{bmatrix}$$

So one example of the infinite solutions is

$$\underline{x} = \begin{bmatrix} 1/3 \\ 1/3 \\ 0 \end{bmatrix}$$

1.8. Eigenanalysis

Eigenanalysis involves the determination of eigenvalues and eigenvectors. It is a part of linear algebra that is extremely important to scientists and engineers in a broad variety of applications. Here we first provide the mathematical framework for obtaining eigenvalues and eigenvectors. Then we provide an example.

For an $n \times n$ square matrix, there are n eigenvalues, though they need not all be different. If the determinant of the matrix is non-zero, all of the eigenvalues are non-zero. If the determinant of the matrix is zero, at least one of the eigenvalues is zero.

To calculate the eigenvalues, $\{\lambda\}$, for an $n \times n$ matrix, one begins by subtracting the eigenvalue from all diagonal elements.

$$\underline{\underline{A}} - \lambda \underline{\underline{I}} = \begin{bmatrix} a_{1,1} - \lambda & a_{1,1} & \cdots & a_{1,n} \\ a_{2,1} & a_{2,2} - \lambda & \cdots & a_{2,n} \\ \vdots & \vdots & \ddots & \vdots \\ a_{n,1} & a_{n,2} & \cdots & a_{n,n} - \lambda \end{bmatrix} \qquad (1.38)$$

Second, the determinant of $\underline{\underline{A}} - \lambda \underline{\underline{I}}$ is set to zero,

$$\det(\underline{\underline{A}} - \lambda \underline{\underline{I}}) = 0 \qquad (1.39)$$

Third you must solve this equation for λ. Equation (1.39) is called the characteristic equation and is a polynomial in λ of order n. Thus, this equation has n roots. As with any polynomial equation, the roots may be complex. The n roots of equation (1.39) are the n eigenvalues.

Each eigenvalue has associated with it an eigenvector. Thus, if there are n-eigenvectors, there are also n-eigenvalues. The ith eigenvector, $\underline{w}_i$, for the matrix $\underline{\underline{A}}$, is obtained by solving

$$(\underline{\underline{A}} - \lambda_i \underline{\underline{I}})\underline{w}_i = \underline{0} \qquad (1.40)$$

for $\underline{w}_i$. This equation defines the eigenvectors and can be solved n times for all n eigenvalues to yield n eigenvectors.

For a 2x2 matrix, we have

$$\underline{\underline{A}}_{2x2} - \lambda \underline{\underline{I}}_{2x2} = \begin{bmatrix} a_{1,1} - \lambda & a_{1,2} \\ a_{2,1} & a_{2,2} - \lambda \end{bmatrix} \tag{1.41}$$

The determinant of $\underline{\underline{A}} - \lambda \underline{\underline{I}}$ is set to zero,

$$\det\left(\underline{\underline{A}}_{2x2} - \lambda \underline{\underline{I}}_{2x2}\right) = (a_{1,1} - \lambda)(a_{2,2} - \lambda) - a_{1,2}a_{2,1} = \lambda^2 - (a_{1,1} + a_{2,2})\lambda + \det\left(\underline{\underline{A}}_{2x2}\right) = 0 \tag{1.42}$$

For a 2x2 matrix, the characteristic equation is a quadratic polynomial. The two roots of the equation are given by the quadratic formula,

$$\lambda = \frac{(a_{1,1} + a_{2,2}) \pm \sqrt{(a_{1,1} + a_{2,2})^2 - 4\det\left(\underline{\underline{A}}_{2x2}\right)}}{2} \tag{1.43}$$

The eigenvectors are given by

$$\left(\underline{\underline{A}}_{2x2} - \lambda_i \underline{\underline{I}}_{2x2}\right) \underline{w}_i = \begin{bmatrix} a_{1,1} - \lambda_i & a_{1,2} \\ a_{2,1} & a_{2,2} - \lambda_i \end{bmatrix} \underline{w}_i = \underline{0} \tag{1.44}$$

This set of linear equations must be solved. The first step in solving a system of linear equation is finding the determinant. Because the eigenvalues are solutions to $\det\left(\underline{\underline{A}} - \lambda \underline{\underline{I}}\right) = 0$ (from equation (1.42)), the determinant of the matrix in equation (1.44) is always zero. Since the b vector in equation (1.44) is the zero vector, the rank of $\left(\underline{\underline{A}} - \lambda \underline{\underline{I}}\right)$ is equal to the rank of the augmented matrix $\left(\underline{\underline{A}} - \lambda \underline{\underline{I}}\right) \underline{b}$, which is less than n,

$$\text{rank}\left[\left(\underline{\underline{A}} - \lambda \underline{\underline{I}}\right)\right] = \text{rank}\left[\left(\underline{\underline{A}} - \lambda \underline{\underline{I}}\right) \underline{b}\right] < n \tag{1.45}$$

Consequently, there are always infinite solutions for each eigenvector. Another way to think of this is that eigenvectors provide directions only, but not magnitude.

For a 2x2 matrix, we can randomly set the second element of the eigenvector to an arbitrary variable, x. Solving the first equation in equation (1.44) yields the eigenvectors,

$$\underline{w}_i = \begin{bmatrix} -a_{1,2}x \\ (a_{1,1} - \lambda_i)x \end{bmatrix} \text{ for } i = 1 \text{ to } n, \text{ for abitrary } x \neq 0 \tag{1.46}$$

Typically, eigenvectors are reported as normalized vectors, where the magnitude of the vector is one. The magnitude of an arbitrary vector, $\underline{x}$, of length n is defined as

$$|\underline{x}| = \sqrt{\sum_{i=1}^{n} x_i^2} \qquad (1.47)$$

Therefore a normalized vector, $\underline{x}'$, is given by

$$\underline{x}' = \frac{1}{|\underline{x}|}\underline{x} \qquad (1.48)$$

By construction, the magnitude of this normalized vector is one. The normalized eigenvectors for the 2x2 example is then

$$\underline{w}'_i = \frac{1}{|\underline{w}_i|}\underline{w}_i = \frac{1}{\sqrt{\left(a_{1,2}^2 + (a_{1,1} - \lambda_i)^2\right)x^2}}\begin{bmatrix} -a_{1,2}x \\ (a_{1,1} - \lambda_i)x \end{bmatrix} \text{ for } i = 1 \text{ to } n, \text{ for abitrary } x \neq 0 \qquad (1.49)$$

Even the normalized eigenvectors still have two equivalent expressions, which involves multiplication by -1. A normalized eigenvector multiplied by -1 is still a normalized eigenvector. For a common list of eigenvectors, we adopt the convention that the real component of the first element of each eigenvector should be positive.

Example 1.4.

Consider the 2x2 matrix, $\underline{\underline{A}} = \begin{bmatrix} 1 & 2 \\ 2 & 1 \end{bmatrix}$. The characteristic equation is given by

$$\det(\underline{\underline{A}} - \lambda \underline{\underline{I}}) = (1 - \lambda)(1 - \lambda) - 4 = \lambda^2 - 2\lambda - 3 = 0$$

The eigenvalues are given by $\lambda_1 = -1$ and $\lambda_2 = 3$. From equation (1.46), the unnormalized eigenvectors are for $x = 1$,

$$\underline{w}_1 = \begin{bmatrix} -2 \cdot 1 \\ (1 - -1) \cdot 1 \end{bmatrix} = \begin{bmatrix} -2 \\ 2 \end{bmatrix} \qquad \underline{w}_2 = \begin{bmatrix} -2 \cdot 1 \\ (1 - 3) \cdot 1 \end{bmatrix} = \begin{bmatrix} -2 \\ -2 \end{bmatrix}$$

The normalized eigenvectors are

$$\underline{w}'_1 = \frac{1}{|\underline{w}_1|}\underline{w}_1 = \frac{1}{2\sqrt{2}}\underline{w}_1 = \begin{bmatrix} -1/\sqrt{2} \\ 1/\sqrt{2} \end{bmatrix} \qquad \underline{w}'_2 = \frac{1}{|\underline{w}_2|}\underline{w}_2 = \frac{1}{2\sqrt{2}}\underline{w}_2 = \begin{bmatrix} -1/\sqrt{2} \\ -1/\sqrt{2} \end{bmatrix}$$

If we follow the convention that the first element should be positive in our eigenvectors, then our normalized eigenvectors are

$$\underline{w}'_1 = \begin{bmatrix} 1/\sqrt{2} \\ -1/\sqrt{2} \end{bmatrix} \quad \underline{w}'_2 = \begin{bmatrix} 1/\sqrt{2} \\ 1/\sqrt{2} \end{bmatrix}$$

Example 1.5. Normal Mode Analysis

Students frequently ask for an example of the physical meaning of eigenvalues and eigenvectors. Such a meaning can be directly observed in the application of normal mode analysis. Below we provide a normal mode analysis of a simple one-dimensional model of carbon dioxide in the ideal gas state.

Consider a carbon dioxide molecule modeled as three particles connected by two springs, with the carbon atom in the middle, as shown in Figure 1.1. The positions in the laboratory frame of reference are subscripted with an "0". The positions as deviations from their equilibrium positions relative to the molecule center-of-mass contain only the essential information for this problem and do not have the "0" subscript.

We model the interaction between molecules as Hookian springs. For a Hookian spring, the potential energy, U, is

$$U = \frac{k}{2}(x - x_0)^2$$

and the force, F, is

$$F = -k(x - x_0)$$

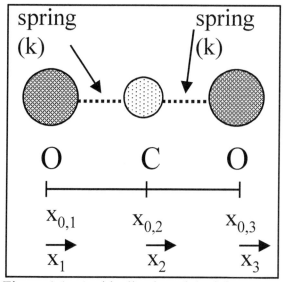

Figure 1.1. An idealized model of the carbon dioxide molecule.

where k is the spring constant (units of kg/s²). We can write Newton's equations of motion for the three molecules:

$$m_O a_1 = F_1 = k(x_2 - x_1)$$
$$m_C a_2 = F_2 = -k(x_2 - x_1) + k(x_3 - x_2)$$
$$m_O a_3 = F_3 = -k(x_3 - x_2)$$

Knowing that the acceleration is the second derivative of the position, we can rewrite the above equations in matrix form as (first divide both side of all of the equations by the masses)

$$\frac{d^2 \underline{x}}{dt^2} = \underline{\underline{A}}\,\underline{x}$$

where

$$\underline{\underline{A}} = \begin{bmatrix} -k/m_O & k/m_O & 0 \\ k/m_C & -2k/m_C & k/m_C \\ 0 & k/m_O & -k/m_O \end{bmatrix} \qquad \underline{x} = \begin{bmatrix} x_1 \\ x_2 \\ x_3 \end{bmatrix}$$

Today, we are not interested in solving this systems of ordinary differential equations. We are content to perform an eigenanalysis on the matrix $\underline{\underline{A}}$. First we have

$$\underline{\underline{A}} - \lambda \underline{\underline{I}} = \begin{bmatrix} -k/m_O - \lambda & k/m_O & 0 \\ k/m_C & -2k/m_C - \lambda & k/m_C \\ 0 & k/m_O & -k/m_O - \lambda \end{bmatrix}$$

The characteristic equation for this matrix is

$$\det(\underline{\underline{A}} - \lambda \underline{\underline{I}}) = \left(-k/m_O - \lambda\right)\left(-2k/m_C - \lambda\right)\left(-k/m_O - \lambda\right) - \frac{k}{m_O}\frac{k}{m_C}\left(-k/m_O - \lambda\right) - \frac{k}{m_O}\frac{k}{m_C}\left(-k/m_O - \lambda\right) = 0$$

Rearranging this cubic polynomial for λ yields

$$\det(\underline{\underline{A}} - \lambda \underline{\underline{I}}) = -\lambda\left(\lambda + \frac{k}{m_O}\right)\left(\lambda + \frac{k}{m_O}\left(1 + 2\frac{m_O}{m_C}\right)\right) = 0$$

In this form, by inspection the roots to the characteristic equation are

$$\lambda_1 = 0 \qquad \lambda_2 = -\frac{k}{m_O} \qquad \lambda_3 = -\frac{k}{m_O}\left(1 + 2\frac{m_O}{m_C}\right)$$

The eigenvectors for each of these eigenvalues are given by

$$(\underline{\underline{A}} - \lambda_i \underline{\underline{I}})\underline{w}_i = \underline{0}$$

$$(\underline{\underline{A}} - \lambda_1 \underline{\underline{I}})\underline{w}_1 = \begin{bmatrix} -k/m_O & k/m_O & 0 \\ k/m_C & -2k/m_C & k/m_C \\ 0 & k/m_O & -k/m_O \end{bmatrix} \underline{w}_1 = \underline{0}$$

Since the equations are not linearly independent, we can remove $w_{3,1}$ as a variable and set it equal to 1. Then our system becomes:

$$\begin{bmatrix} -k/m_O & k/m_O \\ k/m_C & -2k/m_C \end{bmatrix} \begin{bmatrix} w_{1,1} \\ w_{2,1} \end{bmatrix} = \begin{bmatrix} 0 \\ -k/m_C \end{bmatrix}$$

This 2x2 matrix has a non-zero determinant. We can solve it uniquely to yield

$$\begin{bmatrix} w_{1,1} \\ w_{2,1} \end{bmatrix} = \begin{bmatrix} 1 \\ 1 \end{bmatrix}$$

and the eigenvector that corresponds to $\lambda_1 = 0$ is

$$\underline{w}_1 = \begin{bmatrix} 1 \\ 1 \\ 1 \end{bmatrix}$$

To find the second eigenvector, the eigenvector that corresponds to $\lambda_2 = -k/m_O$

$$\underline{\underline{A}} - \lambda_2 \underline{\underline{I}} = \begin{bmatrix} 0 & k/m_O & 0 \\ k/m_C & -2k/m_C + k/m_O & k/m_C \\ 0 & k/m_O & 0 \end{bmatrix}$$

As before, we remove the third equation and remove $w_{3,2}$ as a variable and set it equal to 1. Then our system becomes:

$$\begin{bmatrix} 0 & k/m_O \\ k/m_C & -2k/m_C + k/m_O \end{bmatrix} \begin{bmatrix} w_{1,2} \\ w_{2,2} \end{bmatrix} = \begin{bmatrix} 0 \\ -k/m_C \end{bmatrix}$$

With this new matrix, we can calculate that the determinant is non-zero and the solution is

$$\begin{bmatrix} w_{1,2} \\ w_{2,2} \end{bmatrix} = \begin{bmatrix} -1 \\ 0 \end{bmatrix}$$

and the eigenvector that corresponds to $\lambda_2 = -k/m_O$ is

$$\underline{w}_2 = \begin{bmatrix} -1 \\ 0 \\ 1 \end{bmatrix}$$

To find the third eigenvector, the eigenvector that corresponds to $\lambda_3 = -\dfrac{k}{m_O}\left(1 + 2\dfrac{m_O}{m_C}\right)$

$$\underline{\underline{A}} - \lambda_3 \underline{\underline{I}} = \begin{bmatrix} 2k/m_C & k/m_O & 0 \\ k/m_C & k/m_O & k/m_C \\ 0 & k/m_O & 2k/m_C \end{bmatrix}$$

As before, we remove the third equation and remove $w_{3,3}$ as a variable and set it equal to 1. Then our system becomes:

$$\begin{bmatrix} 2k/m_C & k/m_O \\ k/m_C & k/m_O \end{bmatrix} \begin{bmatrix} w_{1,3} \\ w_{2,3} \end{bmatrix} = \begin{bmatrix} 0 \\ -k/m_C \end{bmatrix}$$

With this new matrix, we can calculate that the determinant is non-zero and the solution is

$$\begin{bmatrix} w_{1,3} \\ w_{2,3} \end{bmatrix} = \begin{bmatrix} 1 \\ -\frac{2m_O}{m_C} \end{bmatrix}$$

and the eigenvector that corresponds to $\lambda_3 = -\frac{k}{m_O}\left(1 + 2\frac{m_O}{m_C}\right)$ is

$$\underline{w}_3 = \begin{bmatrix} 1 \\ -\frac{2m_O}{m_C} \\ 1 \end{bmatrix}$$

So we have the three eigenvalues and the three eigenvectors. So what? What good do they do us? For a vibrating molecule, the square root of the absolute value of the eigenvalues from doing an eigenanalysis of Newton's equations of motion, as we have done, are the normal frequencies. You see that the units of the eigenvalues are 1/sec², so the square root has units of frequency (or inverse time).

For carbon dioxide, the three normal frequencies are:

$$\underline{\omega} = \begin{bmatrix} 0 \\ \sqrt{\frac{k}{m_O}} \\ \sqrt{\frac{k}{m_O}\left(1 + \frac{2m_O}{m_C}\right)} \end{bmatrix}$$

The frequency of zero is no frequency at all. It is not a vibrational mode. In fact, it is a translation of the molecule. We can see this by examining the eigenvectors. The eigenvector that corresponds to $\lambda_1 = 0$ or $\omega_1 = \sqrt{|\lambda_1|} = 0$ is

$$\underline{w}_1 = \begin{bmatrix} 1 \\ 1 \\ 1 \end{bmatrix}$$

This is a description of the normal vibration associated with frequency of zero. It says that all atoms move the same amount in the x-direction. See Figure 1.2.

The eigenvector that corresponds to $\lambda_2 = -k/m_O$ or $\omega_2 = \sqrt{|\lambda_2|} = \sqrt{\dfrac{k}{m_O}}$ is

$$\underline{w}_2 = \begin{bmatrix} -1 \\ 0 \\ 1 \end{bmatrix}$$

This eigenvector describes a vibration where both the oxygen move away from the carbon equally and the carbon does not move.

The eigenvector that corresponds to
$$\lambda_3 = -\dfrac{k}{m_O}\left(1 + 2\dfrac{m_O}{m_C}\right) \text{ or}$$

$$\omega_3 = \sqrt{|\lambda_3|} = \sqrt{\dfrac{k}{m_O}\left(1 + 2\dfrac{m_O}{m_C}\right)} \text{ is}$$

$$\underline{w}_2 = \begin{bmatrix} 1 \\ -\dfrac{2m_O}{m_C} \\ 1 \end{bmatrix}$$

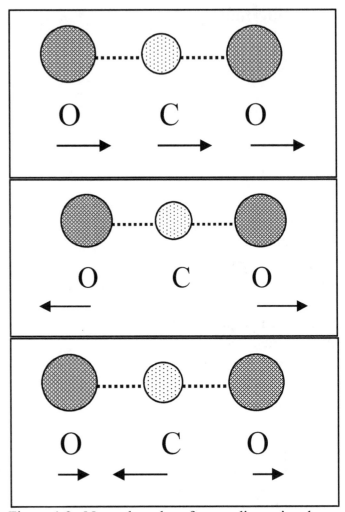

Figure 1.2. Normal modes of a one-dimensional model of carbon dioxide. The top mode is a translational mode with $\omega_1 = 0$. The middle mode is a vibrational mode with $\omega_2 = \sqrt{\dfrac{k}{m_O}}$. The bottom mode is a vibrational mode with $\omega_3 = \sqrt{\dfrac{k}{m_O}\left(1 + 2\dfrac{m_O}{m_C}\right)}$.

This eigenvector describes a vibration where both the oxygen move to the right and the carbon move more to the left, in such a way that there is no center of mass motion.

The normal modes of motion provide a complete, independent set of vibrations from which any other vibration is a linear combination.

1.9. Summary of Logically Equivalent Statements

At this point, we have identified the most common tasks required in the solution of a system of linear algebraic equations. An example of the analytical method by which numerical values can be obtained by hand has been presented. The value in presenting these hand calculations lies in developing an understanding of the general behavior of systems of linear equations. It is unlikely that we will ever be called upon (outside of exams) to calculate eigenvalues or inverses by hand. Nevertheless, knowing what to expect from an analytical understanding better prepares us to make sense of the numerical tools and better understand why numerical tools fail. For example, we can ask a program to compute the inverse of a matrix with a determinant of zero. Depending on the software, a variety of cryptic messages may be provided when the code crashes. Showing that the determinant is zero first, allows us to understand that not all of our equations were independent. Alternatively, some software will simply return some matrix without ever notifying the user that the inverse does not exist. Again, the basic understanding provided above can go a long way in interpreting the results of mathematical software.

To this end, we can identify summarize the logically equivalent statements about an $n \times n$ matrix, $\underline{\underline{A}}$. If any one of these statements is true, all the others are true.

• If and only if $\det(\underline{\underline{A}}) \neq 0$	• If and only if $\det(\underline{\underline{A}}) = 0$
• then inverse exists	• then inverse does not exist
• then $\underline{\underline{A}}$ is non-singular	• then $\underline{\underline{A}}$ is singular
• then $rank(\underline{\underline{A}}) = n$	• then $rank(\underline{\underline{A}}) < n$
• then there are no zero rows in the row echelon form of $\underline{\underline{A}}$	• then there is at least one zero row in the row echelon form of $\underline{\underline{A}}$
• then $\underline{\underline{A}}\underline{x} = \underline{b}$ has one, unique solution	• then $\underline{\underline{A}}\underline{x} = \underline{b}$ has either no solution or infinite solutions
• all eigenvalues of $\underline{\underline{A}}$ are non-zero	• at least one eigenvalue of $\underline{\underline{A}}$ is zero

1.10. Summary of MATLAB Commands

In the table below, a summary of important linear algebra commands in MATLAB is given.

Entering a matrix	

`A=[a11,12;a21,a22]`
 	
 (commas separate elements in a row, semicolons separate rows)
 (easiest for direct data entry)
 	
 `A=[a11     a12`
 `a21     a22]`
 	
 (tabs separate elements in a row, returns separate rows)
 (useful for copying data from a table in Word or Excel)	

Entering a column vector	Entering a row vector
`b=[b1;b2;b3]` (an nx1 vector)	`b=[b1,b2,b3]` (a 1xn vector)
determinant of a matrix `det(A)` (scalar)	rank of a matrix `rank(A)` (scalar)
inverse of an nxn matrix `inv(A)` (nxn matrix)	transpose of an nxm matrix or an nx1 vector `A=A'` (mxn matrix or 1xn vector)
solution of Ax=b `x=A\b` or `x=inv(A)*b` (nx1 vector)	reduced row echelon form of an nxn matrix `rref(A)` (nxn matrix)
eigenvalues and eigenvector of an nxn matrix `[w,lambda]=eig(A)` (w is an nxn matrix where each column is an eigenvector, lambda is a nxn matrix where each diagonal element is an eigenvalue, off-diagonals are zero).	

1.11. Problems

For each of the three problems below, find
(a) the determinant of A
(b) the reduced row echelon form of A
(c) the reduced row echelon form of the Augmented Ab matrix.
(d) the rank of A.
(e) the rank of the augmented Ab matrix.
(f) the inverse of A if it exists
(g) a solution to $\underline{\underline{A}}\underline{x} = \underline{b}$ if it exists.

Problem 1.1.

$$\underline{\underline{A}} = \begin{bmatrix} 1 & 2 & -1 \\ 1 & 5 & 1 \\ 1 & 4 & 2 \end{bmatrix} \text{ and } \underline{b} = \begin{bmatrix} 1 \\ 2 \\ 5 \end{bmatrix}$$

Problem 1.2.

$$\underline{\underline{A}} = \begin{bmatrix} 1 & 1 & 1 & 1 \\ 1 & 2 & 1 & -1 \\ 1 & 3 & 2 & -2 \\ 1 & 4 & 3 & -3 \end{bmatrix} \text{ and } \underline{b} = \begin{bmatrix} 1 \\ 0 \\ 1 \\ 2 \end{bmatrix}$$

Problem 1.3.

$$\underline{\underline{A}} = \begin{bmatrix} 1 & 2 & -1 \\ 1 & 5 & 1 \\ 2 & 7 & 0 \end{bmatrix} \text{ and } \underline{b} = \begin{bmatrix} 1 \\ 2 \\ 5 \end{bmatrix}$$

Chapter 2. Regression

2.1. Introduction

Regression is a term describing the fitting of a particular function to a set of data through optimization of the parameters or constant coefficients that appear in the function. From this point of view regression is a kind of optimization, which is the subject of Chapter 8. However, many common forms of regression involve the solution of a linear set of algebraic equations. Thus, just as we separate our discussion of the solution of linear (Chapter 1) and nonlinear (Chapter 4) algebraic equations, so too do we separate our discussion of linear and nonlinear optimization. We shall call this linear optimization by the term regression and place it directly after the chapter on linear algebra because from one point of view, it is simply an application of linear algebra.

2.2. Single Variable Linear Regression

Imagine that we have a set of n data points (x_i, y_i) for $i = 1$ to n. Perhaps, we have a theory that tells us that y should be a linear function of x. We know that the equation of a line is given by

$$\hat{y} = b_1 x + b_0 \tag{2.1}$$

where b_1 is the slope and b_0 is the y-intercept of the line. The hat on $\hat{y}$ reminds us that this value of y comes from the model and not from data. We would like to know what the best values of the constant coefficients, b_1 and b_0. If we substitute our data points into this equation, it will not in general be satisfied due to noise in the data. Therefore, we create the equality by introducing the error in the ith data point, e_i, such that

$$y_i = b_1 x_i + b_0 + e_i \tag{2.2}$$

The best-fit model will minimize the error. In particular we wish to minimize the Sum of the Square of Errors, *SSE*, defined as

$$SSE \equiv \sum_{i=1}^{n} e_i^2 = \sum_{i=1}^{n} (y_i - \hat{y}_i)^2 = \sum_{i=1}^{n} (y_i - b_1 x_i - b_0)^2 \quad (2.3)$$

In order to minimize the function *SSE* with respect to b_1 and b_0, we take the partial differential of *SSE* with respect to b_1 and b_0 and set them equal to zero. (We remember from differential calculus that the derivative of a function is zero at a minimum or maximum.)

$$\frac{\partial SSE}{\partial b_0} = -2 \sum_{i=1}^{n} (y_i - b_1 x_i - b_0) = 0 \quad (2.4.a)$$

$$\frac{\partial SSE}{\partial b_1} = -2 \sum_{i=1}^{n} (y_i - b_1 x_i - b_0) x_i = 0 \quad (2.4.b)$$

We then solve these two equations for b_1 and b_0. Notice that equations (2.4.a) and (2.4.b) are linear in the unknown variables, b_1 and b_0. We have already seen in Chapter 1, the solution to a set of two linear algebraic equations. In this case, we have

$$b_1 = \frac{\sum_{i=1}^{n}(x_i - \bar{x})(y_i - \bar{y})}{\sum_{i=1}^{n}(x_i - \bar{x})^2} \quad (2.5.a)$$

$$b_0 = \frac{\sum_{i=1}^{n} y_i - b_1 \sum_{i=1}^{n} x_i}{n} = \bar{y} - b_1 \bar{x} \quad (2.5.b)$$

where $\bar{x}$ and $\bar{y}$ are the average values of the set of *x* and *y* respectively.

We note that one requires two points at a minimum to perform a single-variable linear regression. Since there are two parameters, we need two data points. One can see from equation (2.5) that if the two data points have the same value of *x*, the slope is infinite. Therefore, this method requires at least two data points at different values of the independent variable, *x*.

2.3. The Variance of the Regression Coefficients

From a statistical point of view, the regression coefficients in equation (2.5) are mean values of the slope and intercept. Whenever a mean is calculated, a variance can also be calculated. Here provided without derivation is the variance of the regression coefficients. The variance of the slope is

$$\sigma_{b_1}^2 = \frac{\sigma^2}{\sum_{i=1}^{n}(x_i - \bar{x})^2} \tag{2.6.a}$$

The variance of the y-intercept is

$$\sigma_{b_0}^2 = \frac{\sum_{i=1}^{n} x_i^2}{n \sum_{i=1}^{n}(x_i - \bar{x})^2} \sigma^2 \tag{2.6.b}$$

where σ^2 is the model error variance. An unbiased estimate of σ^2 is s^2 where

$$\sigma^2 \approx s^2 = \frac{SSE}{n-2} \tag{2.7}$$

Frequently, one would like a convenient metric to express the goodness of the fit. The measure of fit, *MOF*, provides a way to determine if the best-fit model is a good model. A common definition of the *MOF* is

$$MOF = \frac{SSR}{SST} = 1 - \frac{SSE}{SST} \tag{2.8}$$

where the sum of the squares of the errors, SSE, was defined in equation (2.3), and where the Sum of the Squares of the Regression, SSR, is based on the variance of the model predictions and the average values of *y*.

$$SSR = \sum_{i=1}^{n}(\hat{y}_i - \bar{y})^2 \tag{2.9}$$

The Sum of the Squares of the Total variance, SST, is

$$SST = SSR + SSE = \sum_{i=1}^{n}(\hat{y}_i - \bar{y})^2 + \sum_{i=1}^{n}(y_i - \hat{y}_i)^2 \qquad (2.10)$$

An analysis of equation (2.10) makes it clear that there are two sources of variance, that captured by the regression, SSR, and that outside the regression, SSE. This *MOF* is the fraction of variance captured by the regression. It is bounded between 0 and 1. A value of *MOF* of 1 means the model fits the data perfectly. The farther the value of the MOF is below 1, the worse the fit of the model.

We note in passing that a more rigorous method for evaluating the goodness of the fit is to use the f-distribution. We can define a variable, f, by

$$f = \frac{SSR}{s^2} \qquad (2.11)$$

For a given level of confidence, γ, and for a given number of data points, n, one can determine whether the regression models the data to within that level of confidence by direct comparison with the appropriate critical value of the f statistic

$$f > f_\gamma(1, n-2) \qquad (2.12)$$

where $f_\gamma(v_1, v_2)$ is defined as the lower limit in the following constraint

$$\gamma = \int_{f_\gamma(v_1,v_2)}^{\infty} \frac{\Gamma\left[\dfrac{v_1+v_2}{2}\right]\left(\dfrac{v_1}{v_2}\right)^{\frac{v_1}{2}}}{\Gamma\left[\dfrac{v_1}{2}\right]\Gamma\left[\dfrac{v_2}{2}\right]} \cdot \frac{f^{\frac{v_1}{2}-1}}{\left(1+\dfrac{v_1}{v_2}f\right)^{\frac{v_1+v_2}{2}}} df \qquad (2.13)$$

In equation (2.13), the gamma function, $\Gamma[z]$, is a standard integral

$$\Gamma[z] = \int_0^{\infty} t^{z-1} e^{-t} dt \qquad (2.14.a)$$

For positive integers, the expression simplifies to

$$\Gamma[n] = (n-1) \qquad (2.14.b)$$

In MATLAB, the gamma function can be accessed with the `gamma(z)` command. In MATLAB, the cumulative integral of the f distribution can be accessed with the `fcdf(f,v1,v2)` command. For a given value of f, this function will return $1-\gamma$. Values of the integral in equation (2.13) as a function of γ, v_1 and v_2 are routinely available in tables of critical f values" [Walpole *et al.* 1998].

2.4. Multivariate Linear Regression

Above, we learned how to perform a regression for *y* when it is a linear function of a single variable *x*. In this section, we extend the capability to performing a regression for *y* when it is a linear function of an arbitrary number, *m*, of variables. In this case, our model has the general form

$$\hat{y}_i = b_0 + \sum_{j=1}^{m} b_j x_{i,j} \tag{2.15}$$

Note that there are now two subscripts. The subscript *j* differentiates the different independent variables and runs from 1 to *m*. Perhaps one variable is temperature and another concentration. The subscript *i* designates the individual data points, $(x_{i,1}, x_{i,2} \ldots x_{i,m}, y_i)$, and runs from 1 to *n*. The error for each data point is again defined by the equation

$$y_i = b_0 + \sum_{j=1}^{m} b_j x_{i,j} + e_i \tag{2.16}$$

The method for finding the best-fit parameters for this system is exactly analogous to what we did for the single-variable linear regression. We define, the Sum of the Squares of the Error, *SSE*, exactly as we did before in equation (25.6)

$$SSE \equiv \sum_{i=1}^{n} e_i^2 = \sum_{i=1}^{n} (y_i - \hat{y}_i)^2 = \sum_{i=1}^{n} \left(y_i - b_0 - \sum_{j=1}^{m} b_j x_{i,j} \right)^2 \tag{2.17}$$

We take the partial derivatives of *SSE* with respect to each of the parameters in $\{b\}$ and set them equal to zero. This gives $m+1$ independent, linear algebraic equations of the form:

$$b_0 \sum_{i=1}^{n} x_{i,0} x_{i,j} + b_1 \sum_{i=1}^{n} x_{i,1} x_{i,j} + b_2 \sum_{i=1}^{n} x_{i,2} x_{i,j} + \ldots + b_m \sum_{i=1}^{n} x_{i,m} x_{i,j} = \sum_{i=1}^{n} y_i x_{i,j} \quad \text{for } i = 0 \text{ to } m \tag{2.18}$$

We have introduced a new set of constants, $x_{0,i} = 1$ in this equation to allow for compact expression of the equations. This set of equations can be written in matrix form as

$$\underline{\underline{A}}\,\underline{b} = \underline{g} \tag{2.19}$$

where the j,k element of the matrix $\underline{\underline{A}}$ is defined as

$$A_{j,k} = \sum_{i=1}^{n} x_{i,k-1} x_{i,j-1} \tag{2.20}$$

where the j element of the column vector of constants, $\underline{g}$, is defined as

$$g_j = \sum_{i=1}^{n} y_i x_{i,j-1} \tag{2.21}$$

The solution vector, $\underline{b}$, contains the regression parameters,

$$\underline{b}_j = \begin{bmatrix} b_0 \\ b_1 \\ \vdots \\ b_m \end{bmatrix} \tag{2.22}$$

The presence of the $k-1$ and $j-1$ subscripts is due to the fact that in our model, we numbered our regression parameters starting from 0 rather than 1, but the numbering of rows and columns in the matrix begins at 1.

To determine the variances of the parameters in multivariate linear regression, we use analogous equations as for the single-variable case. The variances of each parameter are defined as

$$\sigma_{b_j}^2 = A_{j-1,j-1}^{-1} \sigma^2 \tag{2.23}$$

where $A_{j-1,j-1}^{-1}$ is the diagonal element of the inverse of $\underline{\underline{A}}$ (which is <u>not</u> the inverse of the corresponding element of $\underline{\underline{A}}$). The covariances are

$$\sigma_{b_j b_k} = A_{j-1,k-1}^{-1} \sigma^2 \tag{2.24}$$

where, as in the single-variable case, an unbiased estimate of σ^2 is given by s^2, which is defined as

$$\sigma^2 \approx s^2 = \frac{SSE}{n-m-2} \tag{2.25}$$

The measure of fit for the multiple regression case is defined in exactly the same way as the single-variable regression case, given in equation (2.8).

2.5. Polynomial Regression

One might think that in a chapter on linear regression, that polynomial regression would be outside the scope of the discussion. However, the defining characteristic of this chapter is that the **parameters in the regression be linear**. The tools in this chapter can be used even if the independent variables can have any sort of nonlinear functionality so long as the regression coefficients appear in a linear form. In a polynomial, the coefficients appear in a linear manner,

$$\hat{y}_i = b_0 + \sum_{j=1}^{m} b_j x_i^j \tag{2.26}$$

We see that the form of the model in a polynomial regression is very similar to the form of our previous multivariate linear regression in equation (2.15).

$$\hat{y}_i = b_0 + \sum_{j=1}^{m} b_j x_{j,i} \tag{2.15}$$

In fact we can perform polynomial regression using the tools from multivariate linear regression if

$$x_{j,i} = x_i^j \tag{2.27}$$

The rest of the procedure for polynomial regression is then exactly the same as multivariate linear regression.

2.6 Linearization of Equations

Frequently we attempt to fit experimental data with a nonlinear theory. Often the parameterization of this data can be accomplished with linear regression because the theory is linear in the parameters and nonlinear in the independent variables. Consider a general expression for the rate of a process,

$$r = k \exp\left(-\frac{E_a}{k_B T}\right) \tag{2.28}$$

in which the rate, r, is a function of the temperature, T, and two constants, the activation energy, E_a, and the pre-exponential factor, k. (k_B is Boltzmann's constant.) Many physical processes obey this functional form including chemical reactions and diffusion.

Frequently, one has rates as a function of temperature and one wishes to determine the activation barrier for the process and the rate constant, k. In its current form, this equation is nonlinear in the parameters. However, taking the natural log of both sides of the equation yields

$$\ln(r) = \ln(k) - \frac{E_a}{k_B T} \tag{2.29}$$

Lo and behold, now the equation is linear of the form

$$\hat{y} = b_1 x + b_0 \tag{2.1}$$

where $\hat{y} = \ln(r)$, $x = \frac{1}{T}$, $b_0 = \ln(k)$ and $b_1 = -\frac{E_a}{k_B}$. This transformation is so common that it is given its own name, the Arrhenius form of the rate equation. A single variable linear regression can be performed on this data, yielding b_0 and b_1. The rate constant and activation energy can be directly obtained from $k = \exp(b_0)$ and $E_a = -k_B b_1$.

2.7. Confidence Intervals

Sometimes, in addition to the mean and standard deviation of the regression coefficients, one would also like to know the uncertainty in the regression as a function of the independent variables. It is probably not surprising that the uncertainty is typically smaller in the middle of the region where data was extracted and larger at the ends. Such a feat can be accomplished through the use of confidence intervals. First, we provide formulae for confidence intervals on the regression coefficients, then we provide confidence intervals as a function of the independent variable.

Typically one is interested in a 90% (or 95% or 99% or in general $CI\%$) confidence interval, which provides a lower bound and upper bound to a range within which you are $CI\%$ confident that the true result lies. Typically, this confidence, CI, is related to a parameter, α through the expression, $CI = 100(1-2\alpha)$. Thus a confidence interval of 90% corresponds to $\alpha = 5\%$. Under the assumption that the observations (data points) are normally and independently distributed, a confidence interval on the slope, b_1, is given by

$$P(\hat{b} - t_{\alpha,n-2}\sqrt{\sigma_b^2} < b < \hat{b} + t_{\alpha,n-2}\sqrt{\sigma_b^2}) = 1 - 2\alpha \qquad (2.30.a)$$

This mathematical equation states that the probability that the slope lies between the lower and upper limit is CI. The confidence interval on the intercept, b_0, is given by

$$P(\hat{a} - t_{\alpha,n-2}\sqrt{\sigma_a^2} < a < \hat{a} + t_{\alpha,n-2}\sqrt{\sigma_a^2}) = 1 - 2\alpha \qquad (2.30.b)$$

The variances that appear in equation (2.30) are the same as those already computed for the slope and intercept via equation (2.6). The number of data points is n. The critical t-statistic, $t_{\alpha,v}$, is defined as the lower limit in the following constraint

$$\alpha = \int_{t_{\alpha,v}}^{\infty} \frac{\Gamma\left[\frac{v+1}{2}\right]}{\Gamma\left[\frac{v}{2}\right]\sqrt{\pi v}} \left(1 + \frac{t^2}{v}\right)^{-\frac{v+1}{2}} dt \qquad (2.31)$$

Values of this integral as a function of α and v are routinely available in tables of critical t values" [Walpole et al. 1998].

The confidence interval at any arbitrary value of the independent variable can be obtained as follows. Under the same assumption that the observations are normally and independently distributed, a $CI = 100(1-2\alpha)$ confidence interval on the regression at a point, x_o, is given by

$$P(\hat{y}(x_o) - t_{\alpha,n-2}\sqrt{\sigma_{y(x_o)}^2} < y(x_o) < \hat{y}(x_o) + t_{\alpha,n-2}\sqrt{\sigma_{y(x_o)}^2}) = 1 - 2\alpha \qquad (2.32)$$

where $\hat{y}(x_o) = b_0 + b_1 x_o$ and where

$$\sigma^2_{y(x_o)} = \left[\frac{1}{n} + \frac{(x_o - \bar{x})^2}{\sum_{i=1}^{n} x_i^2 - \frac{1}{n}\left(\sum_{i=1}^{n} x_i\right)^2} \right] \sigma^2 \qquad (2.33)$$

where σ^2 is the model error variance, as estimated in equation (2.7) This allows you to evaluate the confidence interval at every value of x, giving upper and lower confidence limits that are functions of x.

More on confidence intervals is available in the literature.[Montgomery & Runger, 1999].

2.8. Regression Subroutines

Note that in order to present short codes, the versions of the codes given below makes three sacrifices. First, these codes contain no comments or instructions for use. Second these codes contain no error checking. For example it does not check that the user has provided sufficient data points. Third, these codes take advantage of the most succinct MATLAB commands, such as implicit for-loops, which shortens the code but may make the code difficult to understand.

Therefore, on the course website, two entirely equivalent versions of this code are provided and are titled *code.m* and *code_short.m*. The short version is presented here. The longer version, containing instructions and serving more as a learning tool, is not presented here.

Code 2.1. Single Variable Linear Regression (linreg1_short)

The single variable regression described in Section 2.2., can be accomplished using the following MATLAB code.

```
function [b,bsd,MOF] = linreg1_short(x,y);
n = max(size(x));
b = zeros(2,1);
bsd = zeros(2,1);
yhat = zeros(n,1);
xsum = sum(x);
xavg = xsum/n;
ysum = sum(y);
yavg = ysum/n;
x2sum = sum(x.*x);
x2avg = x2sum/n;
xvarsum = sum((x-xavg).*(x-xavg));
yvarsum = sum((y-yavg).*(y-yavg));
xycovarsum = sum((x-xavg).*(y-yavg));
b(2) = xycovarsum/xvarsum;
b(1) = yavg - b(2)*xavg;
```

```
yhat = b(1) + b(2)*x;
SSR = sum((yhat - yavg).^2);
SSE = sum((y - yhat).^2);
SST = SSR + SSE;
MOF = SSR/SST;
sigma = SSE/(n-2);
bsd(2) = sigma/xvarsum;
bsd(1) = x2avg*sigma/xvarsum;
bsd(1:2) = sqrt(bsd(1:2));
figure(1);
plot(x,y,'ro'), xlabel( 'x' ), ylabel ( 'y' );
hold on;
plot(x,yhat,'k-');
hold off;
```

An example of using linreg1_short is given below.

```
» [b,bsd,MOF] =
linreg1_short([1; 2; 3; 4.5],
[4; 5.1; 6; 7.3])

b =
   3.14672897196262
   0.93457943925234

bsd =
   0.10993530680867
   0.03756962849017

MOF =

   0.99677841217186
```

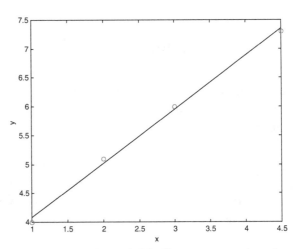

Figure 2.1. Single variable linear regression for Example 2.1.

In this example the mean value of the intercept is 3.147 with a standard deviation of 0.110. The mean value of the slope is 0.934 with a standard deviation of 0.038. The measure of fit is very good at 0.997.

Code 2.2. Multiariate Linear Regression (linregn_short)

The multivariate linear regression described in Section 2.4., can be accomplished using the following MATLAB code.

```
function [b,bsd,MOF] = linregn_short(m,x,y);
n = max(size(y));
mp1 = m + 1;
b = zeros(mp1,1);
bsd = zeros(mp1,1);
yhat = zeros(n,1);
g = zeros(mp1,1);
a = zeros(mp1,mp1);
```

```
xp1 = ones(n,mp1);
xp1(1:n,2:mp1) = x(1:n,1:m);
gvec = [xp1]'*y;
Amat = [xp1]'*xp1;
detA = det(Amat);
Amatinv = inv(Amat);
b = Amatinv*gvec;
dof = n - m;
yhat = xp1*b;
yavg = sum(y)/n;
SSR = sum((yhat - yavg).^2);
SSE = sum((y - yhat).^2);
SST = SSR + SSE;
MOF = SSR/SST;
s2 = SSE/dof;
for j = 1:m
   bsd(j) =sqrt(Amatinv(j,j)*s2);
end
```

An example of using linregn_short is given below.

```
» [b,bsd,MOF] = linregn_short(2,[1 0; 2 5; 3 10; 4 20],[1.9; 13; 24; 45])

b =
    0.79999999999995
    1.11666666666656
    1.98666666666671

bsd =
    0.08660254037844
    0.06972166887784
                   0

MOF =
    0.99999835603242
```

In this example the mean value of the intercept is 0.80 with a standard deviation of 0.087. The mean value of the coefficient for variable 1 is 1.117 with a standard deviation of 0.070. The mean value of the coefficient for variable 2 is 1.987 with a standard deviation of 0.0. The measure of fit is essentially perfect at 1.0.

Code 2.3. Multiariate Linear Regression (polyreg_short)

The polynomial regression described in Section 2.5., can be accomplished using the following MATLAB code.

```
function [b,bsd,MOF] = polyreg(m,x,y);
n = max(size(y));
mp1 = m + 1;
b = zeros(mp1,1);
bsd = zeros(mp1,1);
```

```
yhat = zeros(n,1);
g = zeros(mp1,1);
a = zeros(mp1,mp1);
xp1 = ones(n,mp1);
for i = 2:1:mp1
   xp1(1:n,i) = x(1:n).^(i-1);
end
gvec = [xp1]'*y;
Amat = [xp1]'*xp1;
detA = det(Amat);
Amatinv = inv(Amat);
b = Amatinv*gvec;
dof = n - m;
yhat = xp1*b;
yavg = sum(y)/n;
SSE = 0.0;
SSR = 0.0;
for i = 1:n
   SSR = SSR + (yhat(i) - yavg)^2;
   SSE = SSE + (y(i) - yhat(i))^2;
end
SST = SSR + SSE;
MOF = SSR/SST;
%
s2 = SSE/(dof);
for j = 1:m
   bsd(j) =sqrt(Amatinv(j,j)*s2);
end
figure(1);
plot(x,y,'ro'), xlabel( 'x' ), ylabel ( 'y' );
hold on;
plot(x,yhat,'k-');
hold off;
```

An example of using polyreg_short is given below.

```
» [b,bsd,MOF] = polyreg(2,[1; 2; 3; 4],[4; 11; 22; 36])

b =
   0.24999999999977
   1.94999999999982
   1.74999999999997


bsd =
   0.44017042154147
   0.40155946010522
                  0

MOF =
```

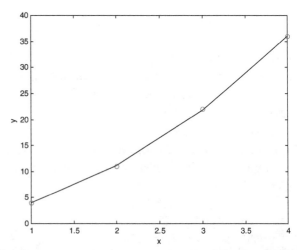

Figure 2.2. Polynomial regression of order 2 for Example 2.3.

0.99991449337324

In this example the mean value of the intercept is 0.25 with a standard deviation of 0.44. The mean value of the linear coefficient is 1.95 with a standard deviation of 0.040. The mean value of the quadratic coefficient is 1.75 with a standard deviation of 0.0. The measure of fit is essentially perfect at 1.0.

Code 2.4. Single Variable Linear Regression with Confidence Intervals (linreg1ci_short)

The single variable regression described in Section 2.2. with confidence intervals described in Section 2.7., can be accomplished using the following MATLAB code.

```
function [b,bsd,bcilo,bcihi,MOF] = linreg1ci(x,y,CI);
n = max(size(x));
b = zeros(2,1);
bsd = zeros(2,1);
bcilo = zeros(2,1);
bcihi = zeros(2,1);
yhat = zeros(n,1);
xsum = sum(x);
xavg = xsum/n;
ysum = sum(y);
yavg = ysum/n;
x2sum = sum(x.*x);
x2avg = x2sum/n;
xvarsum = sum((x-xavg).*(x-xavg));
yvarsum = sum((y-yavg).*(y-yavg));
xycovarsum = sum((x-xavg).*(y-yavg));
b(2) = xycovarsum/xvarsum;
b(1) = yavg - b(2)*xavg;
yhat = b(1) + b(2)*x;
SSR = sum((yhat - yavg).^2);
SSE = sum((y - yhat).^2);
SST = SSR + SSE;
MOF = SSR/SST;
sigma = SSE/(n-2);
bsd(2) = sigma/xvarsum;
bsd(1) = x2avg*sigma/xvarsum;
bsd(1:2) = sqrt(bsd(1:2));
v = n-2;
alpha = (1 - CI/100)/2;
%  need critical value of t statistic here!
cipalpha = CI/100.0 + alpha;
talpha = icdf('t',cipalpha,v);
bcilo(1) = b(1) - talpha*bsd(1);
bcihi(1) = b(1) + talpha*bsd(1);
bcilo(2) = b(2) - talpha*bsd(2);
bcihi(2) = b(2) + talpha*bsd(2);
xmin = min(x);
xmax = max(x);
nx = 20;
```

```
dx = (xmax - xmin)/nx;
xci = zeros(nx,1);
ycilo = zeros(nx,1);
ycihi = zeros(nx,1);
for i = 1:1:nx+1
  xci(i) = xmin + (i-1)*dx;
end
for i = 1:1:nx+1
    thing = 1/n + (xci(i) - xavg)^2/(x2sum - xsum*xsum/n);
    sig = sqrt(thing*sigma);
    ycilo(i) = b(1) + b(2)*xci(i) - talpha*sig;
    ycihi(i) = b(1) + b(2)*xci(i) + talpha*sig;
end
figure(1);
plot(x,y,'ro'), xlabel( 'x' ), ylabel ( 'y' );
hold on;
plot(x,yhat,'k-');
hold on;
plot(xci,ycilo,'g--');
hold on;
plot(xci,ycihi,'g--');
hold off;
```

An example of using linregci_short is given below.

```
» [b,bsd,bcilo,bcihi,MOF] =
linreg1ci_short([1;2;3;4;5;6],[4.2;5.1;6.1;6.7;8.2;8.9],95)

    b =
        3.19333333333333
        0.95428571428571

    bsd =
        0.18429617295288
        0.04732288856688

    bcilo =
        2.16164335714309
        0.68937218408834

    bcihi =
        4.22502330952358
        1.21919924448308

    MOF =
        0.99025920227246
```

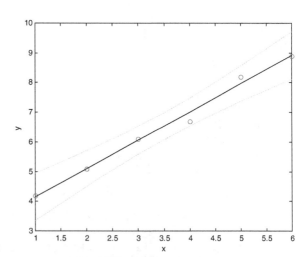

Figure 2.3. Single variable linear regression with 95% confidence intervals for Example 2.4.

In this example, the mean value of the slope is 0.954. There is a 95% probability that the true slope lies between 0.689 and 1.219. The mean value of the intercept is 03.193. There is a 95% probability that the true slope lies between 2.161 and 4.225.

2.9. Problems

For problems 2.1. through 2.4., use the data in Table 2.1. provided at the end of the chapter.

Problem 2.1.
Perform a single-variable linear regression using the model

$$y = b_0 + b_1 x$$

(a) Report the mean value and standard deviation of the regression coefficients.
(b) Report the measure of fit.

Problem 2.2.
Perform a single-variable linear regression using the model

$$y = b_0 + b_1 x + b_2 x^2$$

(a) Report the mean value and standard deviation of the regression coefficients.
(b) Report the measure of fit.

Problem 2.3.
Perform a single-variable linear regression using the model

$$y = b_0 + b_1 x_1 + b_2 x_2$$

(a) Report the mean value and standard deviation of the regression coefficients.
(b) Report the measure of fit.

Problem 2.4.
Consider the isomerization reaction:
$$A \rightarrow B$$
The reaction rate is given by

$$rate = C_A k_o e^{-\frac{E_a}{RT}} \quad \text{[moles/liter/minute]}$$

where
  concentration of A: C_A [moles/liter]

prefactor: k_o [1/minute]
activation energy for reaction: E_a [Joules/mole]
constant: $R = 8.314$ [Joules/mole/K]
temperature: T [K]

Determine the rate constants, k_o and E_a, from experimental data. The reaction is measured at a constant concentration of A, $C_A = 0.1$ mol/liter, over a variety of temperatures. The rate is recorded.

Convert the data into the form necessary for a linear regression.

$$\ln(rate) - \ln(C_A) = -\frac{E_a}{RT} + \ln(k_o)$$

This equation is of the form: $y = b_1 x + b_0$ where

$$y = \ln(rate) - \ln(C_A),\ b_1 = E_a,\ x = -\frac{1}{RT},\ \text{and}\ b_0 = \ln(k_o).$$

Problem 2.1.		Problem 2.2.		Problem 2.3.			Problem 2.4.	
y	x	y	x	y	x1	x2	T (K)	rate (mol/min)
15.10065	1	28.60334	2	-0.9819	2	10	275	24.19187
20.49803	2	61.67478	4	-11.3611	4	10	280	23.66394
25.56137	3	113.3302	6	-17.9244	6	10	285	26.15821
28.32206	4	173.9519	8	-32.461	8	10	290	24.65746
34.77068	5	259.4289	10	-36.2547	10	10	295	25.00509
40.13972	6	364.7832	12	-47.5626	12	10	300	26.94418
42.16968	7	488.4343	14	-57.2561	14	10	305	28.67903
50.76091	8	613.0563	16	-67.026	16	10	310	30.44936
55.67701	9	749.2099	18	-83.6231	18	10	315	27.20014
64.44763	10	929.1701	20	-86.6674	20	10	320	30.5438
67.07516	11	1097.771	22	29.37499	2	20	325	33.03537
69.47986	12	1267.19	24	19.8235	4	20	330	31.15566
72.70851	13	1476.381	26	10.10711	6	20	335	31.27551
75.14606	14	1669.017	28	1.793521	8	20	340	33.91974
80.80169	15	1892.182	30	-14.1845	10	20	345	37.48515
95.33654	16	2295.778	32	-15.3942	12	20	350	33.50134
95.69951	17	2525.079	34	-30.6416	14	20	355	36.64442
95.95821	18	2702.216	36	-41.2898	16	20	360	36.46332
98.30454	19	3189.197	38	-43.6643	18	20	365	38.69183
108.0071	20	3487.682	40	-61.3515	20	20	370	38.05063
110.0038	21	3694.302	42	59.34267	2	30	375	43.16253
112.1662	22	4098.896	44	50.36288	4	30	380	37.25768
127.7834	23	4356.963	46	38.2946	6	30	385	44.58887
131.9088	24	5004.153	48	35.14299	8	30	390	42.75867
130.5765	25	5451.785	50	22.59593	10	30	395	39.63743
128.4932	26	5617.26	52	5.427403	12	30	400	46.8552
133.1759	27	6112.091	54	4.737152	14	30	405	47.6692
162.5784	28	6884.878	56	-16.1513	16	30	410	46.85518
153.0207	29	6744.363	58	-21.1513	18	30	415	50.30944
167.7321	30	7731.94	60	-31.7201	20	30	420	49.31385
152.5957	31	7987.73	62	88.41653	2	40	425	47.42592
161.1343	32	8608.879	64	78.65594	4	40	430	48.23446
164.8263	33	9395.567	66	71.87739	6	40	435	45.17433
164.1293	34	9189.839	68	63.23496	8	40	440	47.77558
197.3908	35	10143.08	70	50.03163	10	40	445	53.03878
174.5867	36	11260.03	72	41.4232	12	40	450	50.81684
184.2806	37	11256.51	74	35.46361	14	40	455	53.80978
209.6302	38	11886.53	76	16.42225	16	40	460	59.39249
213.285	39	12858.59	78	19.30515	18	40	465	49.77457
192.9746	40	13275.05	80	-0.39774	20	40	470	54.68614
211.4016	41	13828.53	82	119.5469	2	50	475	51.76229
225.463	42	14263.1	84	105.7219	4	50	480	62.83728
203.6621	43	15320.2	86	97.52663	6	50	485	62.55782
221.1033	44	15382.82	88	86.32457	8	50	490	61.1546
240.4743	45	16761.52	90	82.54179	10	50	495	63.00538
244.3957	46	17750.05	92	70.49636	12	50	500	65.97716
235.5752	47	18869.43	94	56.31296	14	50	505	63.41012
266.7086	48	19735.41	96	59.31316	16	50	510	63.49422
242.1551	49	19618.65	98	41.8617	18	50	515	59.2734
247.5312	50	20856.26	100	32.60213	20	50	520	65.1637
		22302.75	102	152.736	2	60	525	65.39843
		21780.9	104	140.4219	4	60	530	67.05946
		23643.85	106	128.4831	6	60	535	65.32489
		24454.11	108	124.8209	8	60	540	68.30498
		24092.93	110	110.9015	10	60	545	60.15969
		25493.11	112	99.99487	12	60	550	72.99886
		27720.47	114	95.88626	14	60	555	71.30153
		26540.05	116	76.34477	16	60	560	72.98946

	27569.95	118	72.08759	18	60	565	71.34359
	28286.17	120	58.79183	20	60	570	76.5776
	30893.3	122	184.518	2	70	575	64.02322
	31889.66	124	170.9852	4	70	580	69.63294
	32580.52	126	162.2654	6	70	585	71.56308
	33439.5	128	149.4984	8	70	590	77.33005
	34291.1	130	141.5502	10	70	595	76.53128
	35931.36	132	124.0389	12	70	600	67.78118
	37540.16	134	123.5222	14	70	605	75.55963
	38308.29	136	112.349	16	70	610	68.58622
	37227.47	138	105.6767	18	70	615	73.75325
	39111.73	140	83.05119	20	70	620	70.39462
	39514.13	142	214.6974	2	80	625	78.58048
	42198.54	144	198.0034	4	80	630	69.574
	44411.2	146	189.7924	6	80	635	80.03241
	45407.02	148	174.662	8	80	640	79.78581
	44340.07	150	161.9951	10	80	645	73.95461
	47592.02	152	159.3915	12	80	650	86.75819
	46539.26	154	152.3981	14	80	655	84.66778
	51103.05	156	146.9087	16	80	660	74.3263
	51566.66	158	137.7315	18	80	665	83.97316
	50153.09	160	112.7271	20	80	670	87.45661
	52929.14	162	241.3304	2	90	675	88.40439
	57249.2	164	226.2701	4	90	680	89.77914
	55952.01	166	228.4258	6	90	685	88.63773
	58548.59	168	215.4355	8	90	690	76.95617
	59970.3	170	201.0728	10	90	695	79.64456
	60096.43	172	184.5658	12	90	700	76.8413
	63557.29	174	177.1243	14	90	705	88.41259
	64520.17	176	162.3722	16	90	710	83.60883
	61910.48	178	161.6507	18	90	715	94.04191
	64940.35	180	149.2336	20	90	720	83.48615
	67455.36	182	268.5166	2	100	725	86.46614
	66888.78	184	259.1211	4	100	730	81.23222
	68852.54	186	252.9704	6	100	735	88.55475
	69975.65	188	239.0521	8	100	740	84.90952
	75728.08	190	224.2653	10	100	745	85.61426
	72261.33	192	214.1161	12	100	750	91.20792
	73535.18	194	212.5499	14	100	755	97.16316
	81143.03	196	206.598	16	100	760	87.53307
	83236.72	198	188.826	18	100	765	85.33877
	81985.44	200	181.0279	20	100	770	83.61217
						775	85.88699
						780	90.2451
						785	90.96235
						790	94.88847
						795	102.9028
						800	93.0499
						805	90.60689
						810	94.70386

Table 2.1. Data for problems 2.1. to 2.4.

Chapter 3. Numerical Differentiation

3.1. Introduction

The ability to numerically evaluate derivatives at particular points is important in several applications. Sometimes we have experimental data for concentration as a function of time. However, our models are based on reaction rates, that is the change in concentration as a function of time. Therefore, in order to make a direct comparison between the experimental data and the model, we must either integrate the model or differentiate the data. For the time being, we will postpone judgment on the relative merits of the two options. We shall be content to be able to perform either option as need dictates.

There are also situations in the solution of nonlinear algebraic equations where having derivatives are desirable but the analytical forms of the functions are unavailable. In this case, again, we must rely on numerical differentiation.

3.2. Taylor Series Expansions

Many of the most common formulae for numerical derivatives are derived from a Taylor Series Expansion. As a reminder, a Taylor Series expansion of a function $f(x)$, about a point, x_0, is given by

$$f(x) = f(x_0) + \left.\frac{df}{dx}\right|_{x_0}(x - x_0) + \frac{1}{2!}\left.\frac{d^2 f}{dx^2}\right|_{x_0}(x - x_0)^2 + \ldots + \frac{1}{n!}\left.\frac{d^n f}{dx^n}\right|_{x_0}(x - x_0)^n + \ldots \qquad (3.1.a)$$

or equivalently as

$$f(x) = \sum_{n=0}^{\infty} \frac{1}{n!}\left.\frac{d^n f}{dx^n}\right|_{x_0}(x - x_0)^n \qquad (3.1.b)$$

Frequently, we assume that the x-axis is discretized into points. The step size, $(x_{i+1} - x_i)$, is replaced with the variable h, leaving in which case the Taylor series expansion can be written as

$$f(x_{i+1}) = f(x_i) + \left.\frac{df}{dx}\right|_{x_i} h + \frac{1}{2!}\left.\frac{d^2 f}{dx^2}\right|_{x_i} h^2 + \ldots + \frac{1}{n!}\left.\frac{d^n f}{dx^n}\right|_{x_i} h^n + \ldots \qquad (3.1.c)$$

In practice the Taylor Series is truncated after a particular term so that one has an error of the order of first missing term,

$$f(x_{i+1}) = \sum_{n=0}^{m} \frac{1}{n!}\left.\frac{d^n f}{dx^n}\right|_{x_i} h^n + O(h^{m+1}) \qquad (3.1.d)$$

3.3. Finite Difference Formulae

Taylor Series are used to derive formula for numerical derivatives. There are an infinite number of formulae. We will derive a few common formula. Remember as we do this that we can derive equations for different levels of derivative, the first derivative, the second derivative etc. We can also derive equations with different orders of error, depending upon where we truncate the series. Longer (higher order) series provide more accurate estimates of the derivative. Third, we can choose in which direction to expand the Taylor series, forward (in the positive x direction), backward (in the negative x direction) or centered (in both positive and negative x directions). These three choices lead to a myriad of formula; for example one can have a second order numerical formula for the first derivative in the forward direction.

We begin with first derivatives. If we truncate the Taylor Series at the first order term we have

$$f(x_{i+1}) = f(x_i) + \left.\frac{df}{dx}\right|_{x_i} h + O(h^2) \qquad (3.2)$$

Solving for the derivative yields

$$\left.\frac{df}{dx}\right|_{x_i} = \frac{f(x_{i+1}) - f(x_i)}{h} + O(h) \qquad (3.3)$$

where it can be shown that the truncated term is now of order 1. Thus equation (3.3) provides the first-order forward finite difference formula for the first derivative.

Differentiation - 53

If we take the Taylor series in the negative x-direction, we have an analogous expression

$$f(x_{i-1}) = f(x_i) - \left.\frac{df}{dx}\right|_{x_i} h + O(h^2) \tag{3.4}$$

Solving for the derivative yields

$$\left.\frac{df}{dx}\right|_{x_i} = \frac{f(x_i) - f(x_{i-1})}{h} + O(h) \tag{3.5}$$

Thus we have the first-order backward finite difference formula for the first derivative. If we subtract equation (3.4) from equation (3.2) we obtain

$$f(x_{i+1}) - f(x_{i-1}) = 2\left.\frac{df}{dx}\right|_{x_i} h + O(h^3) \tag{3.6}$$

The error in this expression is of order h^3 because the $O(h^2)$ terms cancelled. Rearranging for the derivative yields

$$\left.\frac{df}{dx}\right|_{x_i} = \frac{f(x_{i+1}) - f(x_{i-1})}{2h} + O(h^2) \tag{3.7}$$

Thus we have a second order centered finite difference formula for the first derivative. In all of these expressions we estimate the value of the first derivative based on function evaluations alone.

This general procedure for using the Taylor Series to generate estimates for derivatives can be applied to generate higher order approximations for higher order derivatives. We shall simply state that the lowest order formulae for the second derivatives are for the forward finite difference,

$$\left.\frac{d^2 f}{dx^2}\right|_{x_i} = \frac{f(x_{i+2}) - 2f(x_{i+1}) + f(x_i)}{h^2} + O(h) \tag{3.8.a}$$

for the reverse finite difference,

$$\left.\frac{d^2 f}{dx^2}\right|_{x_i} = \frac{f(x_i) - 2f(x_{i-1}) + f(x_{i-2})}{h^2} + O(h) \tag{3.8.b}$$

and for the centered finite difference,

$$\left.\frac{d^2 f}{dx^2}\right|_{x_i} = \frac{f(x_{i+1}) - 2f(x_i) + f(x_{i-1})}{h^2} + O(h^2) \tag{3.8.c}$$

The next higher order expressions for the first derivatives are for the forward finite difference,

$$\left.\frac{df}{dx}\right|_{x_i} = \frac{-f(x_{i+2}) + 4f(x_{i+1}) - 3f(x_i)}{2h} + O(h^2) \tag{3.9.a}$$

for the reverse finite difference,

$$\left.\frac{df}{dx}\right|_{x_i} = \frac{3f(x_i) - 4f(x_{i-1}) + f(x_{i-2})}{2h} + O(h^2) \tag{3.9.b}$$

and for the centered finite difference,

$$\left.\frac{df}{dx}\right|_{x_i} = \frac{-f(x_{i+2}) + 8f(x_{i+1}) - 8f(x_{i-1}) + f(x_{i-2})}{12h} + O(h^4) \tag{3.9.c}$$

The next higher order expressions for the second derivatives are for the forward finite difference,

$$\left.\frac{d^2 f}{dx^2}\right|_{x_i} = \frac{-f(x_{i+3}) + 4f(x_{i+2}) - 5f(x_{i+1}) + 2f(x_i)}{h^2} + O(h^2) \tag{3.10.a}$$

for the reverse finite difference,

$$\left.\frac{d^2 f}{dx^2}\right|_{x_i} = \frac{2f(x_i) - 5f(x_{i-1}) + 4f(x_{i-2}) - f(x_{i-3})}{h^2} + O(h^2) \tag{3.10.b}$$

and for the centered finite difference,

$$\left.\frac{d^2 f}{dx^2}\right|_{x_i} = \frac{-f(x_{i+2}) + 16f(x_{i+1}) - 30f(x_i) + 16f(x_{i-1}) - f(x_{i-2})}{12h^2} + O(h^4) \tag{3.10.c}$$

Formulae for higher order derivatives are available elsewhere.[Chapra & Canale, 1988.]

3.4. Approximations for Partial Derivatives

Finite difference formulae exist for partial derivatives of functions of more than one variable. We simply report a few of the most useful formulae here. The expression for the first derivatives do not change in form. For example the centered finite difference formula for the ordinary differential given in equation (3.7)

$$\left.\frac{df}{dx}\right|_{x_i} = \frac{f(x_{i+1}) - f(x_{i-1})}{2h} + O(h^2) \tag{3.7}$$

becomes for partial derivative with respect to variable x_j

$$\left.\left(\frac{df}{dx_j}\right)\right|_{x_{k \neq j}}\Bigg|_{x_j^{(i)}} = \frac{f(x_1^{(i)} \ldots x_j^{(i+1)} \ldots x_n^{(i)}) - f(x_1^{(i)} \ldots x_j^{(i-1)} \ldots x_n^{(i)})}{2h_j} + O(h_j^2) \tag{3.11}$$

Let it be clear that the j subscript on the variable x indicates a different independent variable. The superscript (i) on the x is not an exponent. The parentheses are included to make clear it is a notation that does not signify a mathematical operation. Instead, the superscript (i) indicates a different value of x, which we shall soon see can be associated with the i-1, i and i+1 nodes. The subscript that now appears outside the parentheses, $x_{k \neq j}$, indicates that all variables except x_j are held constant in the differentiation. The final subscript $x_j^{(i)}$ of the partial derivative indicates the value of x_j where the derivative is evaluated.

The second partial derivative with respect to variable x_j is analogous to the centered finite difference formula for the ordinary second differential given in equation (3.8.c)

$$\left.\frac{d^2 f}{dx^2}\right|_{x_i} = \frac{f(x_{i+1}) - 2f(x_i) + f(x_{i-1})}{h^2} + O(h^2) \tag{3.8.c}$$

becomes

$$\left.\left(\frac{\partial^2 f}{\partial x_j^2}\right)\right|_{x_{k \neq j}}\Bigg|_{x_j^{(i)}} = \frac{f(x_1^{(i)} \ldots x_j^{(i+1)} \ldots x_n^{(i)}) - 2f(x_1^{(i)} \ldots x_j^{(i)} \ldots x_n^{(i)}) + f(x_1^{(i)} \ldots x_j^{(i-1)} \ldots x_n^{(i)})}{h_j^2} + O(h_j^2) \tag{3.12}$$

Finally, the mixed second partial derivative with respect to variables x_j and x_m is

$$\left(\frac{\partial}{\partial x_m}\left(\frac{\partial f}{\partial x_j}\right)_{x_{k\neq j}}\right)_{x_{k\neq m}}\Bigg|_{x_j^{(i)},x_m^{(i)}} = \frac{\begin{bmatrix} f\left(x_1^{(i)}\ldots x_j^{(i+1)},x_m^{(i+1)}\ldots x_n^{(i)}\right)-f\left(x_1^{(i)}\ldots x_j^{(i+1)},x_m^{(i-1)}\ldots x_n^{(i)}\right) \\ -f\left(x_1^{(i)}\ldots x_j^{(i-1)},x_m^{(i+1)}\ldots x_n^{(i)}\right)+f\left(x_1^{(i)}\ldots x_j^{(i-1)},x_m^{(i-1)}\ldots x_n^{(i)}\right) \end{bmatrix}}{4h_j h_m}+O(h_j h_m)$$

(3.13)

which is symmetric with respect to the order of differentiation,

$$\left(\frac{\partial}{\partial x_m}\left(\frac{\partial f}{\partial x_j}\right)_{x_{k\neq j}}\right)_{x_{k\neq m}}\Bigg|_{x_j^{(i)},x_m^{(i)}} = \left(\frac{\partial}{\partial x_j}\left(\frac{\partial f}{\partial x_m}\right)_{x_{k\neq m}}\right)_{x_{k\neq j}}\Bigg|_{x_m^{(i)},x_j^{(i)}}$$

(3.14)

3.5. Noise

Life is noisy. There is always uncertainty associated with the measurements of data. Nowhere is the impact of noise in the data more evident than in numerical differentiation. To illustrate this point, we shall apply some of the formulae presented in the previous section and observe their accuracy as a function of the amount of noise in the data.

We shall investigate a toy problem, with a simple monotonically increasing function, $f(x) = x^2 + x^3$, along a small range, from 0 to 1. The function is shown without any noise in Figure 3.1. It is plotted with a data discretization of 0.01 in the x-direction. Also shown is the function with varying degrees of noise ranging from 0.01% to 20% noise. The noise is randomly generated at each data point.

We shall investigate the effect of noise on centered-finite difference formulae, although the conceptual

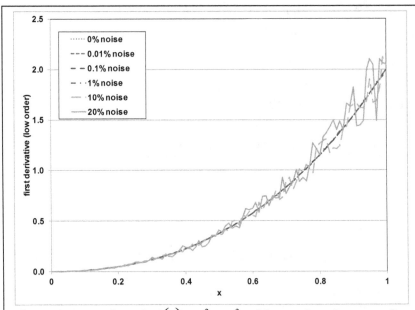

Figure 3.1. A plot of $f(x) = x^2 + x^3$ with varying degrees of noise in the data.

results is the same for the forward and backward finite difference formulae. We presented four centered finite difference formula in the previous section: (i) a second-order approximation to the first derivative, equation (3.7), (ii) a fourth-order approximation to the first derivative, equation (3.9.c), (iii) a second order approximation to the second derivative, equation (3.8.c) and (iv) a fourth order approximation to the second derivative, equation (3.10.c). In Figure 3.2, we present two plots for each of the four finite difference formulae. The first plot is of the derivative itself. The second plot is presented on a semi-log plot and presents the absolute value of the relative error of the derivative referenced to the analytical value of the derivative.

We begin the discussion by noting that the function in Figure 3.1 retains the correct shape, despite the presence of noise. In practice we may judge this data to be fairly good, especially for the lower percentages of noise. However, the derivatives plotted on the left side of Figure 3.2 tell a very different story. What instantly stands out is that the derivatives can easily have the wrong sign, giving the impression that the function is not monotonically increasing, nor always of positive concavity. For all four approximations, we observe from the error plotted on a log axis on the right side that that the error of the approximate derivative increases as the noise increases. In this error plot, a value of 1 corresponds to 100% error. We note that the higher order methods do not demonstrate an appreciable improvement over the lower order methods for noisy data.

Also of note, we observe for the second-order approximation of the first derivative, that there is an error in the derivative even based on the original function with no noise in the range from 10^{-3} to 10^{-5}. This is a measure of the accuracy of the formulae. The polynomial is third order and the formulae is second order so the approximation is not exact. However, when we apply the fourth-order approximation to the first derivative, we find that the error drops to 10^{-16}. The finite difference formuala should be exact. Our calculations only include 16 digits. This error is due to "truncation error" due to the finite precision of our software.

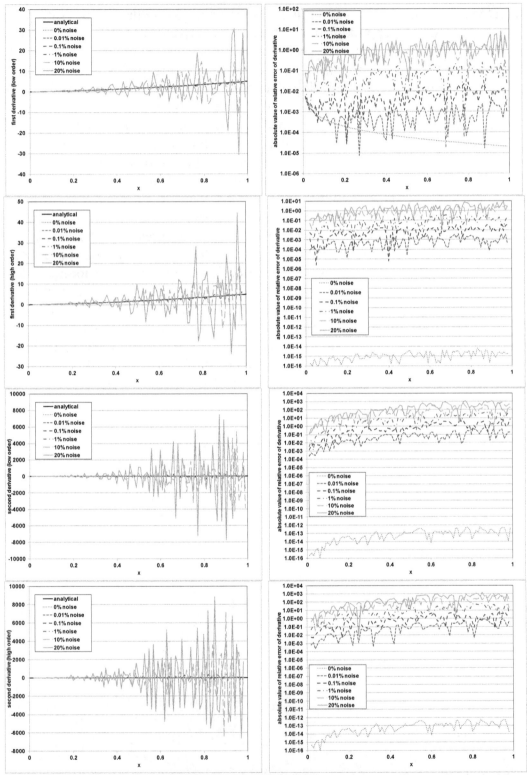

Figure 3.2. Affect of noise on numerical differentiation of centered finite difference formulae.

3.6. Problems

Problem 3.1.
Scour the internet or published literature for experimental data of the form $y(x)$ vs x. Numerically differentiate this data. Plot it. Comment on whether the derivative appears to be reasonable.

Chapter 4. Solution of a Single Nonlinear Algebraic Equation

4.1. Introduction

Life, my friends, is nonlinear. As such, in our roles as problem-solvers, we will be called upon time and again to tackle nonlinear problems. As luck would have it, life hasn't changed much in its nonlinearity from what it was for previous generations. Consequently, the mathematicians and computer scientists that have gone before us have left us with a plethora of tools for subduing all manner of nonlinear beasts. Therefore as we embrace what is undoubtedly a daunting challenge, we do so with the knowledge that, armed with the appropriate tools, the solution of nonlinear algebraic equations too is a task we can methodically add to our repertoire.

The solution of nonlinear algebraic equations is frequently called root-finding, since our goal in this chapter is to find the value of x such that $f(x) = 0$ no matter how pernicious the function may appear. The roots of an equation are those values of x that satisfy $f(x) = 0$.

There are a myriad of way to find roots. The organization of this chapter is to provide a general understanding of the approach to solving nonlinear algebraic equations. We then examine two intuitive approaches that we shall find are deeply flawed. We then examine other approaches that offer some improvements.

4.2. Iterative Solutions and Convergence

Our goal is to find the value of x that satisfies the following equation.

$$f(x) = 0 \tag{4.1}$$

where $f(x)$ is some nonlinear algebraic equation.

All root-finding techniques are iterative, meaning we make a guess and we keep updating our guess based on the value of $f(x)$. We stop updating our guess when we meet some specified convergence criterion. The convergence criteria can be on x or on $f(x)$. One example of a convergence criteria is the absolute value of the function in question, $f(x)$.

$$err_{f,i} = |f(x_i)| \tag{4.2}$$

Since $f(x)$ goes to zero at the root, the absolute value of $f(x)$ is an indication of how close we are to the root, at least in the neighborhood of the root. The problem with this error is that it has units of $f(x)$. If the function is an energy balance, the units of the equation might be kJ/mole or it might be J/mole, which would make the value of $f(x)$ one thousand times larger. If one possesses sufficient familiarity with the problem that one has an absolute measure of tolerance, then one can proceed with this convergence criterion.

Alternatively we can specify a convergence criteria on x itself. In this case, there are two types of convergence criteria, absolute and relative. The absolute error is given by

$$err_{x,i} = |x_i - x_{i-1}| \tag{4.3}$$

Again, this error has units of x. If we want a relative error, which is dimensionless, then we use:

$$err_{x,i} = \left| \frac{x_i - x_{i-1}}{x_i} \right| \tag{4.4}$$

This gives us a percent error based on our current value of x. The advantage of this error is that if we pick a tolerance of 10^{-n}, then our answer will always have n good significant digits. For example, if we use this relative error on x and set our acceptable tolerance to 10^{-6}, then our answer will always have 6 good significant digits. For this reason, the relative error on x is the preferred error to use. However, this error has one drawback that the others do not have. Namely, this error will diverge if the root is located at zero.

Regardless of what choice of error we use, we have to specify a tolerance. The tolerance tells us the maximum allowable error. We stop the iterations when the error is less than the tolerance.

4.3. Successive Approximations

A primitive technique to solve for the roots of $f(x)$ is called successive approximation. In successive approximation, we rearrange the equation so that we isolate x on left-hand side. So for the example $f(x)$ given below

$$f(x) = x - \exp(-x) = 0 \tag{4.5.a}$$

we rearrange as

$$x = \exp(-x) \tag{4.5.b}$$

We then make an initial guess for x, plug it into the right hand side and see if it equals our guess. If it does not, we take the new value of the right-hand side of the equation and use that for x. We continue until our guess gives us the same answer.

Example 4.1. Successive Approximations

Let's find the root to the nonlinear algebraic equation given in equation (4.5.b) using successive approximations. We will use an initial guess of 0.5. We will use a relative error on x as the criterion for convergence and we will set our tolerance at 10^{-6}.

iteration	x	exp(-x)	relative error
1	0.5000000	0.6065307	-
2	0.6065307	0.5452392	1.1241E-01
3	0.5452392	0.5797031	5.9451E-02
4	0.5797031	0.5600646	3.5065E-02
5	0.5600646	0.5711721	1.9447E-02
6	0.5711721	0.5648629	1.1169E-02
7	0.5648629	0.5684380	6.2893E-03
8	0.5684380	0.5664095	3.5815E-03
9	0.5664095	0.5675596	2.0265E-03
10	0.5675596	0.5669072	1.1508E-03
11	0.5669072	0.5672772	6.5221E-04
12	0.5672772	0.5670674	3.7005E-04
13	0.5670674	0.5671864	2.0982E-04
14	0.5671864	0.5671189	1.1902E-04
15	0.5671189	0.5671571	6.7494E-05
16	0.5671571	0.5671354	3.8280E-05
17	0.5671354	0.5671477	2.1710E-05
18	0.5671477	0.5671408	1.2313E-05
19	0.5671408	0.5671447	6.9830E-06
20	0.5671447	0.5671425	3.9604E-06
21	0.5671425	0.5671438	2.2461E-06
22	0.5671438	0.5671430	1.2739E-06
23	0.5671430	0.5671434	7.2246E-07

In this example it took 23 iterations, or evaluations of the function in order to converge (obtain an error less than our specified tolerance). So, we now have converged to a final answer of $x = 0.567143$.

Example 4.2. Successive Approximations

Now let's try to solve an analogous problem. Equation (4.6) has the same root as equation (4.5).

$$f(x) = x + \ln(x) = 0 \qquad (4.6.a)$$

We rearrange the function so as to isolate x on the left-hand side as

$$x = -\ln(x) \qquad (4.6.b)$$

We perform the iterative successive approximation procedure as before. We use the same initial guess of 0.5.

iteration	x	-ln(x)	relative error
1	0.5000000	0.6931472	-
2	0.6931472	0.3665129	8.9119E-01
3	0.3665129	1.0037220	6.3485E-01
4	1.0037220	-0.0037146	2.7121E+02
5	-0.0037146	Does Not Exist	

By iteration 5, we see that we are trying to take the natural log of a negative number, which does not exist. The program crashes. Taken together, these two examples illustrate several key points about successive approximations, which are summarized in the table below.

Successive Approximation	
Advantages	• simple to understand and use
Disadvantages	• no guarantee of convergence • very slow convergence • need a good initial guess for convergence

My advice is to never use successive approximations. As a root-finding method it is completely unreliable. The only reason it is presented here is to try to convince you that you should not use it, no matter how simple it looks.

A MATLAB code which implements successive approximation is provided later in this chapter.

4.4. Bisection Method of Rootfinding

Another method for finding roots is called the bisection method. In the bisection method we still want to find the root to $f(x) = 0$. We do so by finding a value of x, namely x_+, where the function is positive, $f(x) > 0$, and a second value of x, namely x_-, where the function is negative, $f(x) < 0$. These two values of x are called brackets. If we have brackets and our function is continuous, then we know that at least one value of x for which $f(x) = 0$ lies somewhere between the two brackets.

In the bisection method, we initiate the procedure by finding the brackets. The bisection method does not provide brackets. Rather it requires them as inputs. Perhaps, we plot the function and visually identify points where the function is positive and negative. After we have the brackets, we then find the value of x midway between the brackets.

$$x_{mid} = \frac{x_+ + x_-}{2} \tag{4.7}$$

At each iteration, we evaluate the function at the current midpoint. If $f(x_{mid}) > 0$, then we replace x_+ with x_{mid}, namely $x_+ = x_{mid}$. The other possibility is that $f(x_{mid}) < 0$, in which case $x_- = x_{mid}$. With our new brackets, we find the new midpoint and continue the iterative procedure until we have reached the desired tolerance.

Example 4.3. Bisection Method
Let's solve the problem that the successive approximations problem could not solve.

$$f(x) = x + \ln(x) = 0 \tag{4.6.a}$$

We will take as our brackets,

$$x_- = 0.1 \text{ where } f(x_-) = -2.203 < 0$$
$$x_+ = 1.0 \text{ where } f(x_+) = 1.0 > 0$$

How did we find these brackets? It was either by trial and error or we plotted $f(x)$ vs x to obtain some idea where the function was positive and negative.

We will again use a relative error on x as the criterion for convergence and we will set our tolerance at 10^{-6}.

	x_-	x_+	$f(x_-)$	$f(x_+)$	error
1	5.500000E-01	1.000000E+00	-4.783700E-02	1.000000E+00	4.500000E-01
2	5.500000E-01	7.750000E-01	-4.783700E-02	5.201078E-01	2.903226E-01
3	5.500000E-01	6.625000E-01	-4.783700E-02	2.507653E-01	1.698113E-01
4	5.500000E-01	6.062500E-01	-4.783700E-02	1.057872E-01	9.278351E-02
5	5.500000E-01	5.781250E-01	-4.783700E-02	3.015983E-02	4.864865E-02
6	5.640625E-01	5.781250E-01	-8.527718E-03	3.015983E-02	2.432432E-02
7	5.640625E-01	5.710938E-01	-8.527718E-03	1.089185E-02	1.231190E-02
8	5.640625E-01	5.675781E-01	-8.527718E-03	1.201251E-03	6.194081E-03
9	5.658203E-01	5.675781E-01	-3.658408E-03	1.201251E-03	3.097041E-03
10	5.666992E-01	5.675781E-01	-1.227376E-03	1.201251E-03	1.548520E-03
11	5.671387E-01	5.675781E-01	-1.276207E-05	1.201251E-03	7.742602E-04

12	5.671387E-01	5.673584E-01	-1.276207E-05	5.943195E-04	3.872800E-04
13	5.671387E-01	5.672485E-01	-1.276207E-05	2.907975E-04	1.936775E-04
14	5.671387E-01	5.671936E-01	-1.276207E-05	1.390224E-04	9.684813E-05
15	5.671387E-01	5.671661E-01	-1.276207E-05	6.313133E-05	4.842641E-05
16	5.671387E-01	5.671524E-01	-1.276207E-05	2.518492E-05	2.421379E-05
17	5.671387E-01	5.671455E-01	-1.276207E-05	6.211497E-06	1.210704E-05
18	5.671421E-01	5.671455E-01	-3.275270E-06	6.211497E-06	6.053521E-06
19	5.671421E-01	5.671438E-01	-3.275270E-06	1.468118E-06	3.026770E-06
20	5.671430E-01	5.671438E-01	-9.035750E-07	1.468118E-06	1.513385E-06
21	5.671430E-01	5.671434E-01	-9.035750E-07	2.822717E-07	7.566930E-07

After 21 iterations, we have converged to a final answer of x = 0.567143. The bisection method converged even for the form of the equation where successive approximations would not. In fact, the bisection method is guaranteed to converge, if you can first find brackets. However, the bisection method was still pretty slow; it took a lot of iterations.

This example illustrates several key points about the bisection method:

Bisection Method	
Advantages	<ul><li>simple to understand and use</li><li>guaranteed convergence, if you can find brackets</li></ul>
Disadvantages	<ul><li>must first find brackets (i.e., you need a good initial guess of where the solution is)</li><li>very slow convergence</li></ul>

A MATLAB code which implements the bisection method is provided later in this chapter.

4.5. Single Variable Newton-Raphson

One of the most useful root-finding techniques is called the Newton-Raphson method. Like all the methods in this chapter, the Newton-Raphson technique allows you to find solutions to a general non-linear algebraic equation, $f(x) = 0$.

The advantage of the Newton-Raphson method lies in the fact that we include information about the derivative in the iterative procedure. We can approximate the derivative of $f(x)$ at a point x_1 numerically through a finite difference formula,

$$f'(x_1) = \left.\frac{df}{dx}\right|_{x_1} \approx \frac{f(x_1) - f(x_2)}{x_1 - x_2} \tag{4.8}$$

In the Newton-Raphson procedure, we make an initial guess of the root, say x_1. Since we are looking for a root to $f(x)$, let's say that we want x_2 to be a solution to $f(x) = 0$. Let's rearrange the equation to solve for x_2.

$$x_2 = x_1 - \frac{f(x_1) - f(x_2)}{f'(x_1)} \tag{4.9}$$

Now, if x_2 is a solution to $f(x) = 0$, then $f(x_2) = 0$ and the equation becomes:

$$x_2 = x_1 - \frac{f(x_1)}{f'(x_1)} \tag{4.10}$$

This is the Newton-Raphson Method. Based on the value of the function, $f(x_1)$, and its derivative, $f'(x_1)$, at x_1 we estimate the root to be at x_2. Of course, this is just an estimate. The root will not actually be at x_2 (unless the problem is linear). Therefore, we can implement the Newton-Raphson Method as an iterative procedure

$$x_{i+1} = x_i - \frac{f(x_i)}{f'(x_i)} \tag{4.11}$$

until the difference between x_{i+1} and x_i is small enough to satisfy us.

The Newton-Raphson method requires you to calculate the first derivative of the equation, $f'(x)$. Sometimes this may be problematic. Additionally, we see from the equation above that when the derivative is zero, the Newton-Raphson method fails, because we divide by the derivative. This is a weakness of the method.

However because we go to the trouble to give the Newton-Raphson method the extra

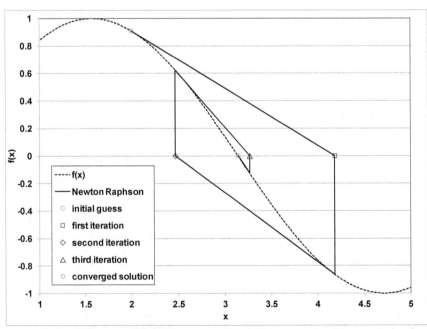

Figure 4.1. Graphical illustration of the Newton-Raphson method.

information about the function contained in the derivative, it will converge must faster than the previous methods. We will see this demonstrated in the following example.

Finally, as with any root-finding method, the Newton-Raphson method requires a good initial guess of the root.

In Figure 4.1., a graphical illustration of the Newton-Raphson method is provided to find the root of $f(x) = \sin(x)$ at $x=\pi$ from an initial guess of $x=2$. One follows the slope at $x=2$ until one reaches, $f(x) = 0$, which specifies the new iterated value of x. The new iteration begins at the point $(x, f(x))$ and this procedure continues until the solution converges within an acceptable tolerance.

Example. 4.4. The Newton-Raphson Method

Let's again solve the problem that the successive approximations problem could not solve.

$$f(x) = x + \ln(x) = 0 \qquad (4.6.a)$$

The derivative is

$$f'(x) = 1 + \frac{1}{x} \qquad (4.12)$$

We will use the same initial guess of 0.5. We will again use a relative error on x as the criterion for convergence and we will set our tolerance at 10^{-6}. At each iteration we must evaluate both the function and its derivative.

```
    x_old           f(x_old)         f'(x_old)        x_new          error
1   5.000000E-01   -1.931472E-01    3.000000E+00    5.643824E-01    -
2   5.643824E-01   -7.640861E-03    2.771848E+00    5.671390E-01    4.860527E-03
3   5.671390E-01   -1.188933E-05    2.763236E+00    5.671433E-01    7.586591E-06
4   5.671433E-01   -2.877842E-11    2.763223E+00    5.671433E-01    1.836358E-11
```

We see in this example that the Newton-Raphson method converged to a root with 11 good significant figures in only four iterations. We observe that for the last three iterations, the error dropped quadratically. By that we mean

$$err_{i+1} \approx err_i^2 \quad \text{or} \quad \frac{err_{i+1}}{err_i^2} \approx 1 \qquad (4.13)$$

Quadratic convergence is a rule of thumb for the Newton-Raphson method. The ratio ought to be on the order of 1. In this example, we have

$$\frac{err_3}{err_2^2} = \frac{7.6 \cdot 10^{-6}}{(4.9 \cdot 10^{-3})^2} = 0.32 \quad \text{and} \quad \frac{err_4}{err_3^2} = \frac{1.83 \cdot 10^{-11}}{(7.6 \cdot 10^{-6})^2} = 0.32$$

We only obtain quadratic convergence near the root. How near do we have to be? The answer to this question depends upon the equation we are solving. It is worth noting that just because the Newton-Raphson method converges to a root from an initial guess, it may not converge to the same root (or converge at all) from all initial guesses closer to the root.

This example illustrates several key points about successive approximations:

Newton-Raphson Method	
Advantages	<ul><li>simple to understand and use</li><li>quadratic (fast) convergence, near the root</li></ul>
Disadvantages	<ul><li>have to calculate analytical form of derivative</li><li>blows up when derivative is zero.</li><li>need a good initial guess for convergence</li></ul>

A MATLAB code which implements the Newton-Raphson method is provided later in this chapter.

4.6. Newton-Raphson with Numerical Derivatives

For whatever reason, people don't like to take derivatives. Therefore, they don't want to use the Newton-Raphson method, since it requires both the function and its derivative. However, we can avoid analytical differentiation of the function through the use of numerical differentiation. For example, we might choose to approximate the derivative at x_i using the second-order centered finite difference formula, as provided in equation (3.7)

$$f'(x_i) = \frac{f(x_i + h) - f(x_i - h)}{2h} \tag{3.7}$$

where h is some small number. Generally I define h according to a rule of thumb

$$h = \min(0.01 \cdot x_i, 0.01) \tag{4.14}$$

This is just a rule of thumb that I made up that seems to work 95% of the time. More sophisticated rules for estimated the increment size certainly exist. Using this rule of thumb, we execute the Newton-Raphson algorithm in precisely the same way, except we never to have to evaluate the derivative analytically.

Example 4.5. Newton-Raphson with Numerical Derivatives

Let's again solve the problem that the successive approximations problem could not solve.

$$f(x) = x + \ln(x) = 0 \qquad (4.6.a)$$

We will use an initial guess of 0.5. We will use a relative error on x as the criterion for convergence and we will set our tolerance at 10^{-6}.

```
   x_old              f(x_old)          f'(x_old)         x_new              error
1  5.000000E-01      -1.931472E-01      3.000067E+00      5.643810E-01       1.000000E+02
2  5.643810E-01      -7.644827E-03      2.771912E+00      5.671389E-01       4.862939E-03
3  5.671389E-01      -1.206407E-05      2.763295E+00      5.671433E-01       7.697929E-06
4  5.671433E-01      -2.862443E-10      2.763282E+00      5.671433E-01       1.826496E-10
```

So we converged to 0.5671433 in only four iterations, just as it did in the rigorous Newton-Raphson method. This example illustrates several key points about successive approximations:

Newton-Raphson with Numerical Derivatives	
Advantages	• simple to understand and use
	• quadratic (fast) convergence, near the root
Disadvantages	• blows up when derivative is zero.
	• need a good initial guess for convergence

A MATLAB code which implements the Newton-Raphson method with numerical derivatives is provided later in this chapter.

4.7. Solution in MATLAB

MATLAB has an intrinsic function to find the root of a single nonlinear algebraic equation. The routine is called *fzero.m*. You can access help on it by typing *help fzero* at the MATLAB command line prompt. You can also access the fzero.m file itself and examine the code line by line. The routine uses a procedure that is classified as a "search and interpolation" technique. The first part of the algorithm requires an initial guess. From this guess, the code searches until it finds two brackets, just as in the bisection case. The function is positive at one bracket and negative at the other. Once the brackets have been obtained, rather than testing the midpoint, as in the bisection method, this routine performs a linear interpolation between the two brackets.

$$x_{new} = x_- - \frac{f(x_-)}{f(x_-) - f(x_+)}(x_- - x_+) \tag{4.15}$$

One of the brackets is replaced with x_{new}, based on the sign of $f(x_{new})$. This procedure is iterated until convergence to the desired tolerance. The actual MATLAB code is a little more sophisticated but we now understand the gist behind a "search and interpolate" method.

The simplest syntax for using the fzero.m code is to type at the command line prompt:

```
>> x = fzero('f(x)',x0,tol,trace)
```

where $f(x)$ is the function we want the roots of, x_0 is the initial guess, and *tol* is the relative tolerance on x., and a non-zero value of *trace* requests iteration information.

Example. 4.6. fzero.m

Let's again solve the problem that the successive approximations problem could not solve.

$$f(x) = x + \ln(x) = 0 \tag{4.6.a}$$

The command at the MATLAB prompt is

```
» x = fzero('x+log(x)',0.5,1.e-6,1)
```

The code output is given below.

```
Func evals        x              f(x)            Procedure
1               0.5           -0.193147          initial
2               0.485858      -0.235981          search
3               0.514142      -0.151113          search
4               0.48          -0.253969          search
5               0.52          -0.133926          search
6               0.471716      -0.279663          search
7               0.528284      -0.109836          search
8               0.46          -0.316529          search
9               0.54          -0.0761861         search
10              0.443431      -0.369781          search
11              0.556569      -0.0293964         search
12              0.42          -0.447501          search
13              0.58           0.0352728         search

Looking for a zero in the interval [0.42, 0.58]

14              0.56831        0.00322159        interpolation
15              0.567143       8.92955e-008      interpolation
16              0.567141      -5.43716e-006      interpolation

x =    0.56714332272548
```

MATLAB's fzero.m (search and interpolate)	
Advantages	• comes with MATLAB • slow convergence
Disadvantages	• has to find brackets before it can begin converging • need a good initial guess for convergence • somewhat difficult to use for more complex problems.

4.8. Existence and Uniqueness of Solutions

When dealing with linear algebraic equations, we could determine how many roots were possible. There were only three choices. Either there was 0, 1, or an infinite number of solutions. When dealing with nonlinear equations, we have no such theory. A nonlinear equation can have 0, 1, 2... up through an infinite number of roots. There is no sure way to tell except by plotting it out. In Figure 4.2., four examples of nonlinear equations with 0, 1, 2 and an infinite number of roots are plotted.

It is important to remember that when you use any of the numerical root-finding techniques described above, you will only find one root at a time. Which root you locate depends upon your choice of method and the initial guess.

Example. 4.7. van der Waals equation of state

When one is interested in finding the several roots of an equation, one must provide multiple different guesses to find each root. To illustrate this problem, we examine the van der Waals equation of state (EOS), which relates pressure, p, to molar volume, V, and temperature, T, via

$$p = \frac{RT}{V-b} - \frac{a}{V^2} \qquad (4.16)$$

where R is the gas constant and a and b are species-dependent van der Waals constants. In truth the van der Waals EOS is a cubic equation of state and can be expressed as a cubic polynomial, which can be exploited to reveal the roots through numerical polynomial solvers. (Interestingly, some software solve for the roots of polynomials by constructing a matrix whose characteristic equation corresponds to the polynomial and then using a routine intended to determine eigenvalues in order to determine the roots of the polynomial.) Here, however, we will not take advantage of this fact but will deal with the EOS in the form presented in equation (4.16).

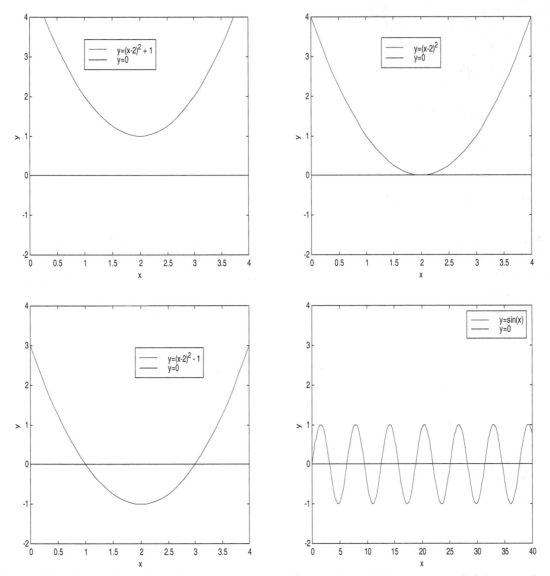

Figure 4.2. Examples of nonlinear equations with zero (top left), one (top right), two (bottom left) and infinite (bottom right) real roots.

At temperature below the critical temperature, the van der Waals EOS predicts vapor-liquid equilibrium. Our task is to find the molar volumes corresponding to the liquid phase, the vapor phase and a third intermediate molar volume that is useful for identifying the vapor pressure at a given temperature. Thus our task is to find all three roots corresponding to a given temperature and pressure.

We rearrange the equation into the familiar form corresponding to $f(x) = 0$

$$f(V) = \frac{RT}{V-b} - \frac{a}{V^2} - p = 0 \tag{4.17}$$

We accept that we are more likely to find relevant roots if we provide initial guesses close to those roots. Therefore it is to our advantage to plot the equation to get a general idea of where the roots lie. For our purposes, let us define a state at $T=98$ K and $p=101325$ Pa. The van der Waals constants for argon are $a=0.1381$ m^6/mol^2 and $b=3.184 \times 10^{-5}$ m^3/mol. The gas constant is $R=8.314$ J/mol/K. The plot is shown in Figure 4.3. Note that the since the three roots occur at different orders of magnitude, the x-axis is

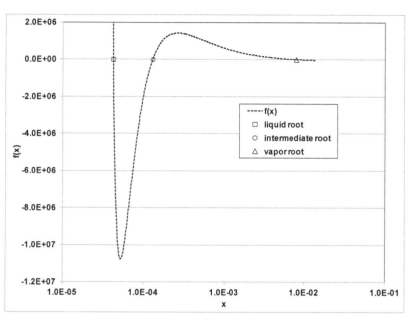

Figure 4.3. van der Waals equation of state (equation (4.17)) plotted showing roots. Note that the x-axis is logarithmic.

plotted on a logarithmic scale. If it were plotted on a linear scale, you wouldn't be able to see the roots. Note also that no values of x are plotted below b. It requires some knowledge of the thermodynamics to realize that the molar volume cannot be less that the van der Waals parameter b. There may be additional mathematical roots that lie below b but they are of no interest to us.

We require good initial guesses in order to find each root. Perhaps the ideal gas law provides an initial guess for the vapor root. The ideal gas law, $pV = RT$, provides an initial guess of $V = 8.041 \times 10^{-3}$ m^3/mol. If we put this in the Newton-Raphson method, we converge in a few iterations to $V_{vapor} = 7.901 \times 10^{-3}$ m^3/mol. So our vapor guess was pretty good and we have found a root. Presumably it is the vapor root, which must have the largest molar volume.

Knowing that the van der Waals parameter b provides a lower limit for the molar volume suggests that we guess something a little larger than b as an initial guess for the liquid root. If we make a guess of $1.1b$ (3.50×10^{-5} m^3/mol), we converge in nine iterations to $V_{liquid} = 4.25 \times 10^{-5}$ m^3/mol.

The intermediate root must lie between these other two roots. Some initial guesses between them will lead to the vapor root and some will lead to the liquid root. There is a range of initial guesses that will lead to the intermediate root. This range is generally defined by the values of x, where the slope points toward the intermediate root. Looking at Figure 4.3 suggests this range may be fairly narrow. If one investigates a range of initial guesses, one finds that initial guesses of

$9x10^{-3} < V < 1.9x10^{-4}$ lead to an intermediate root of $V_{\text{inter}} = 1.29x10^{-4}$ m³/mol. Guesses outside that range lead to other roots including unphysical roots less than b or else the procedure diverges entirely.

This example illustrates a key point in the solution of nonlinear algebraic equations. **Finding an initial guess is often the most important and most difficult part of the problem.** As shown in this example, the good initial guesses came from our understanding of the physical system—that the constant b provided a lower asymptote for the liquid molar volume and that the ideal gas law provided a good estimate of the vapor molar volume.

People who do not rely on numerical methods to solve problems have in the past voiced the criticism that "numerical methods" deprive the student of developing the capability to make reasonable back of the envelope calculations. This example offers the contrary point of view. The ability to solve this problem required reasonable estimates of the roots to initiate the iterative procedure.

4.9. Rootfinding Subroutines

In this section, we provide short routines for implementing four of these root-finding techniques in MATLAB. Note that these codes correspond to the theory and notation exactly as laid out in this book. These codes do not contain extensive error checking, which would complicate the coding and defeat their purpose as learning tools. That said, these codes work and can be used to solve problems.

As before, on the course website, two entirely equivalent versions of this code are provided and are titled *code.m* and *code_short.m*. The short version is presented here. The longer version, containing instructions and serving more as a learning tool, is not presented here. The numerical mechanics of the two versions of the code are identical.

Code 4.1. Successive Approximation (succapp_short)

```
function [x0,err] = succapp_short(x0);
maxit = 100;
tol = 1.0e-6;
err = 100.0;
icount = 0;
xold = x0;
while (err > tol & icount <= maxit)
   icount = icount + 1;
   xnew = funkeval(xold);
   if (icount > 1)
      err = abs((xnew - xold)/xnew);
   end
   fprintf(1,'i = %i xnew = %e xold = %e err = %e \n',icount, xnew, xold, err);
   xold = xnew;
```

Single Nonlinear Algebraic Equation - 75

```
end
x0 = xnew;
if (icount >= maxit)
   fprintf(1,'Sorry.  You did not converge in %i iterations.\n',maxit);
   fprintf(1,'The final value of x was %e \n', x0);
end

function x = funkeval(x0)
x = exp(-x0);
```

An example of using succapp_short is given below.

```
»   [x0,err] = succapp_short(0.5)
i = 1 xnew = 6.065307e-001 xold = 5.000000e-001 err = 1.000000e+002
...
i = 23 xnew = 5.671434e-001 xold = 5.671430e-001 err = 7.224647e-007

x0 =    0.5671
err =   7.2246e-007
```

The root is at 0.5671 and the error is less than the tolerance of 10^{-6}..

Code 4.2. Bisection (bisect_short)

```
function [x0,err] = bisect(xn,xp);
maxit = 100;
tol = 1.0e-6;
err = 100.0;
icount = 0;
fn = funkeval(xn);
fp = funkeval(xp);
while (err > tol & icount <= maxit)
   icount = icount + 1;
   xmid = (xn + xp)/2;
   fmid = funkeval(xmid);
   if (fmid > 0)
      fp = fmid;
      xp = xmid;
   else
      fn = fmid;
      xn = xmid;
   end
   err = abs((xp - xn)/xp);
   fprintf(1,'i = %i xn = %e xp = %e fn = %e fp = %e err = %e \n',icount, xn, xp, fn, fp, err);
end
x0 = xmid;
if (icount >= maxit)
   fprintf(1,'Sorry.  You did not converge in %i iterations.\n',maxit);
   fprintf(1,'The final value of x was %e \n', x0);
end
```

```
function f = funkeval(x)
f = x + log(x);
```

An example of using bisect_short is given below.

```
» [x0,err] = bisect_short(0.1,1.0)
i = 1 xn = 5.500000e-001 xp = 1.000000e+000 fn = -4.783700e-002 fp =
1.000000e+000 err = 4.500000e-001
...
i = 21 xn = 5.671430e-001 xp = 5.671434e-001 fn = -9.035750e-007 fp =
2.822717e-007 err = 7.566930e-007

x0 = 0.5671
err = 7.5669e-007
```

The root is at 0.5671 and the error is less than the tolerance of 10^{-6}..

Code 4.3. Newton-Raphson (newraph1_short)

```
function [x0,err] = newraph1_short(x0);
maxit = 100;
tol = 1.0e-6;
err = 100.0;
icount = 0;
xold =x0;
while (err > tol & icount <= maxit)
   icount = icount + 1;
   f = funkeval(xold);
   df = dfunkeval(xold);
   xnew = xold - f/df;
   if (icount > 1)
      err = abs((xnew - xold)/xnew);
   end
   fprintf(1,'icount = %i xold = %e f = %e df = %e xnew = %e  err = %e
\n',icount, xold, f, df, xnew, err);
   xold = xnew;
end
x0 = xnew;
if (icount >= maxit)
   % you ran out of iterations
   fprintf(1,'Sorry.  You did not converge in %i iterations.\n',maxit);
   fprintf(1,'The final value of x was %e \n', x0);
end

function f = funkeval(x)
f = x + log(x);

function df = dfunkeval(x)
df = 1 + 1/x;
```

Single Nonlinear Algebraic Equation - 77

An example of using newraph1_short is given below.

```
» [x0,err] = newraph1_short(0.5)
icount = 1 xold = 5.000000e-001 f = -1.931472e-001 df = 3.000000e+000 xnew = 5.643824e-001   err = 1.000000e+002
...
icount = 4 xold = 5.671433e-001 f = -2.877842e-011 df = 2.763223e+000 xnew = 5.671433e-001   err = 1.836358e-011

x0 = 0.5671
err = 1.8364e-011
```

The root is at 0.5671 and the error is less than the tolerance of 10^{-6}.

Code 4.4. Newton-Raphson with Numerical derivatives (newraph_short)

```
function [x0,err] = nrnd1(x0);
maxit = 100;
tol = 1.0e-6;
err = 100.0;
icount = 0;
xold =x0;
while (err > tol & icount <= maxit)
   icount = icount + 1;
   f = funkeval(xold);
   h = min(0.01*xold,0.01);
   df = dfunkeval(xold,h);
   xnew = xold - f/df;
   if (icount > 1)
      err = abs((xnew - xold)/xnew);
   end
   fprintf(1,'icount = %i xold = %e f = %e df = %e xnew = %e   err = %e \n',icount, xold, f, df, xnew, err);
   xold = xnew;
end
x0 = xnew;
if (icount >= maxit)
   fprintf(1,'Sorry.  You did not converge in %i iterations.\n',maxit);
   fprintf(1,'The final value of x was %e \n', x0);
end

function f = funkeval(x)
f = x + log(x);

function df = dfunkeval(x,h)
fp = funkeval(x+h);
fn = funkeval(x-h);
df = (fp - fn)/(2*h);
```

An example of using nrnd1_short is given below.

```
» [x0,err] = nrnd1_short(0.5)
icount = 1  xold = 5.000000e-001  f = -1.931472e-001  df = 3.000067e+000  xnew =
5.643810e-001   err = 1.000000e+002
...
icount = 4  xold = 5.671433e-001  f = -2.862443e-010  df = 2.763282e+000  xnew =
5.671433e-001   err = 1.826496e-010

x0 = 0.5671
err = 1.8265e-010
```

The root is at 0.5671 and the error is less than the tolerance of 10^{-6}.

4.10. Problems

Problem 4.1.

Consider the Peng-Robinson Equation of state as given below. The critical temperature and pressure of oxygen are also given below. Each root of this equation is a molar volume. Find all of the roots of the Peng-Robinson equation for oxygen at the temperatures given below and for a pressure of 1.0 atmosphere.

    (a) T = 98.0 K
    (b) T = 298.0 K

(For those of you who have not had thermodynamics, there can only be one phase above the critical temperature.)

$$P = \frac{RT}{V-b} - \frac{a(T)}{V(V+b)+b(V-b)}$$

where, $R = 8.314$ J/mol/K, and $a(T) = 0.45724 \frac{R^2 T_c^2}{P_c} \alpha(T)$, $b = 0.07780 \frac{RT_c}{P_c}$,

$\alpha(T) = \left[1 + \kappa\left(1 - \sqrt{\frac{T}{T_c}}\right)\right]^2$ where, for the oxygen molecule $\kappa = 0.4069$, $T_c = 154.6$ K, and

$P_c = 5.046 \cdot 10^6$ Pa.

Problem 4.2.

Consider the van der Waals equation of state. Vapor-liquid equilibrium at a given temperature, T, occurs at a specific pressure, the vapor pressure, p_{vap}. Therefore, we can consider the vapor pressure our unknown. However, you can't calculate the vapor pressure without knowing the molar volumes of the vapor, V_{vap}, and liquid, V_{liq}, phases. Therefore, we have three unknowns: p_{vap}, V_{vap} and V_{liq}. Solving a system of three equations and three unknowns is much harder than solving a system of equations with one equation and one unknown. Therefore, a useful trick in this problem is to recognize that the van der Waal's equation of state is a cubic equation of state and the roots of a cubic equation of state are easily obtained using the Matlab "roots" function. Therefore, we can pose the solution of the vapor-liquid equilibrium problem of the van der Waals fluid as a single equation with a single equation, in which we guess the vapor pressure, use the roots command to solve for the V_{vap} and V_{liq}. and substitute them into the equation stating that the chemical potentials are equal.

For your reference, the pressure of the van der Waal's equation of state is

$$p = \frac{RT}{V-b} - \frac{a}{V^2}$$

This is a cubic equation of state and can also be written as

$$pV^3 - (pb + RT)V^2 + aV - ab = 0$$

The chemical potential for a van der Waals gas is

$$\mu = -RT\left[\ln\left(\frac{V-b}{\Lambda^3}\right) - \frac{b}{V-b} + \frac{2a}{VRT}\right]$$

This expression for the chemical potential of a van der Waals gas introduces a new constant, the thermal de Broglie wavelength, Λ, but we don't need it because it drops out when we equate chemical potentials.

$$\mu_{liq} - \mu_{vap} = -RT\left[\ln\left(\frac{V_{liq}-b}{\Lambda^3}\right) - \frac{b}{V_{liq}-b} + \frac{2a}{V_{liq}RT}\right] + RT\left[\ln\left(\frac{V_{vap}-b}{\Lambda^3}\right) - \frac{b}{V_{vap}-b} + \frac{2a}{V_{vap}RT}\right]$$

$$= -RT\left[\ln(V_{liq}-b) - \frac{b}{V_{liq}-b} + \frac{2a}{V_{liq}RT}\right] + RT\left[\ln(V_{vap}-b) - \frac{b}{V_{vap}-b} + \frac{2a}{V_{vap}RT}\right]$$

So, in order to solve this problem as a single nonlinear algebraic equation with a single variable, p_{vap}, we have an equation,

$$f(p_{vap}) = \mu_{liq} - \mu_{vap} = 0$$

where we solve for the molar volumes each iteration.

Find the vapor pressure of Argon at $T=77$ K. The van der Waals constants for argon are $a=0.1381$ m^6/mol^2 and $b=3.184 \times 10^{-5}$ m^3/mol. The gas constant is $R=8.314$ J/mol/K. Also report the liquid and vapor molar volumes.

Chapter 5. Solution of a System of Nonlinear Algebraic Equations

5.1. Introduction

Not only is life nonlinear, but its variegated phenomena are typically coupled to each other. The dynamic evolution of a system or a material cannot be described by independent solutions of a material balance, a momentum balance and an energy balance because each equation depends upon the variables solved for in the other equations. Thus we end up with systems of nonlinear equations to describe interesting phenomena. Fear not! Just as was the case in the solution of single nonlinear algebraic equations, today there exist reliable tools to methodically solve systems of nonlinear algebraic equations.

As illustrated in the previous chapter, the challenge in solving a single nonlinear algebraic equation is finding a reasonable initial guess. If you think mucking around in one-dimensional space, looking for an initial guess that will allow the method to converge, is painful, then you can imagine that wandering in n-dimensional space looking for an n-dimensional starting point is significantly more difficult. Nevertheless, it can be done. What is required is access to the appropriate numerical tool coupled with the understanding of the physical system that is provided by the other courses in the undergraduate curriculum.

5.2. Multivariate Newton-Raphson Method

Not surprisingly, the Multivariate Newton-Raphson method is a direct extension of the single variable Newton-Raphson method. Where the single variable Newton-Raphson method solved $f(x) = 0$, the multivariate version will solve a system of n equations of the form

$$f_1(x_1, x_2, x_3, \ldots x_{n-1}, x_n) = 0$$
$$f_2(x_1, x_2, x_3, \ldots x_{n-1}, x_n) = 0$$
$$f_3(x_1, x_2, x_3, \ldots x_{n-1}, x_n) = 0$$
$$\ldots$$
$$f_{n-1}(x_1, x_2, x_3, \ldots x_{n-1}, x_n) = 0$$
$$f_n(x_1, x_2, x_3, \ldots x_{n-1}, x_n) = 0 \tag{5.1}$$

We will adopt the short-hand notation for equation (5.1)

$$\underline{f}(\underline{x}) = 0 \tag{5.2}$$

Note that this short hand notation, which was used for linear algebra, does not here imply anything about the linearity of any of the equations in $\underline{f}(\underline{x})$.

The basis of the single variable Newton-Raphson method lay in the fact that we approximate the derivative of $f(x)$ numerically using a forward finite difference formula based on a truncated Taylor series,

$$f'(x_1) = \left.\frac{df}{dx}\right|_{x_1} \approx \frac{f(x_1) - f(x_2)}{x_1 - x_2} \tag{4.8}$$

Although they were not presented in Chapter 3, multivariate Taylor series also exist. The idea behind the a multivariate Taylor series lies in the definition of the total derivate of a multivariate function. For a function of two variable we can write,

$$df(x_1, x_2) = \left(\frac{\partial f}{\partial x_1}\right)_{x_2} dx_1 + \left(\frac{\partial f}{\partial x_2}\right)_{x_1} dx_2 \tag{5.3}$$

where $\left(\frac{\partial f}{\partial x_1}\right)_{x_2}$ is called the partial derivative of the function, f, with respect to variable, x_1. The subscript outside the parentheses in a partial derivative indicates variables that were treated as constants during the differentiation. For a function of n variables, we have

$$df(\underline{x}) = \sum_{i=1}^{n} \left(\frac{\partial f}{\partial x_i}\right)_{x_{m \neq i}} dx_i \tag{5.4}$$

The multivariate Taylor series expansion, truncated after the first derivative is thus

$$f\left(\underline{x}^{(k)}\right) - f\left(\underline{x}^{(k+1)}\right) = \sum_{i=1}^{n} \left(\frac{\partial f}{\partial x_i}\right)_{x_{m \neq i}} \bigg|_{\underline{x}^{(k)}} \left(x_i^{(k)} - x_i^{(k+1)}\right) \tag{5.5}$$

Let it be clear that the i subscript on the variable x indicates a different independent variable. The superscript (k) on the x is not an exponent. The parentheses are included to make clear it is a notation that does not signify a mathematical operation. Instead, the superscript (k) indicates a different value of x, which we shall soon see can be associated with the k and $k+1$ iterations of the Newton-Raphson method. The subscript that now appears outside the parentheses, $x_{m \neq i}$, indicates that all variables except x_m are held constant in the differentiation. The final subscript $\underline{x}^{(k)}$ of the partial derivative indicates the value of $\underline{x}$ where the derivative is evaluated.

If n=1, then equation (5.5) simplifies to

$$f\left(x_1^{(k)}\right) - f\left(x_1^{(k+1)}\right) = \frac{df}{dx_1}\bigg|_{\underline{x}^{(k)}} \left(x_1^{(k)} - x_1^{(k+1)}\right) \tag{5.6}$$

Equation (5.6) can be rearranged for $x_1^{(k+1)}$

$$x_1^{(k+1)} = x_1^{(k)} - \frac{f\left(x_1^{(k)}\right) - f\left(x_1^{(k+1)}\right)}{\dfrac{df}{dx_1}\bigg|_{x_1^{(k)}}} \tag{5.7}$$

which is precisely the Newton-Raphson method provided in equation (4.11) once we set $f\left(x_1^{(k+1)}\right) = 0$. Similarly for the multivariate case, in which we have n equation and n unknowns, we write equation (5.5) for every function

$$f_j\left(\underline{x}^{(k)}\right) - f_j\left(\underline{x}^{(k+1)}\right) = \sum_{i=1}^{n} \left(\frac{\partial f}{\partial x_i}\right)_{x_{m \neq i}} \bigg|_{\underline{x}^{(k)}} \left(x_i^{(k)} - x_i^{(k+1)}\right) \tag{5.8}$$

where all that has been done is to add a subscript j to f., identifying each equation from $j = 1$ to n. Equation (5.8) is a system of nonlinear algebraic equations. The n unknowns are the next iteration of x, $x_1^{(k+1)}$. Following the Newton-Raphson procedure, we set $f_j\left(\underline{x}^{(k+1)}\right) = 0$ for all j. Again, this choice is based on the fact that we intend for our next estimate to be a better approximation of the root, at which the functions are zero. By convention, we express equation (5.8) in matrix notation as

$$\underline{\underline{J}}^{(k)} \underline{\delta x}^{(k)} = -\underline{R}^{(k)} \tag{5.9}$$

where $\underline{R}^{(k)}$ is called the residual vector at the kth iteration and is defined as

$$\underline{R}^{(k)} \equiv \underline{f}\left(\underline{x}^{(k)}\right) \tag{5.10}$$

In other words, the residual is simply a vector of the values of the functions evaluated at the current guess. The Jacobian matrix at the kth iteration, $\underline{\underline{J}}^{(k)}$, defined as

$$J_{j,i}^{(k)} \equiv \left(\frac{\partial f_j}{\partial x_i}\right)_{x_{m \neq i}}\bigg|_{\underline{x}^{(k)}} \tag{5.11}$$

Thus the Jacobian matrix is an $n \times n$ matrix of all the possible pairwise combinations of partial first derivatives between n unknown variables, x_i, and n functions, f_j. The difference vector, $\underline{\delta x}^{(k)}$, is defined as

$$\underline{\delta x}^{(k)} \equiv \underline{x}^{(k+1)} - \underline{x}^{(k)} \tag{5.12}$$

The difference vector can be rearranged for the value of $\underline{x}$ at the new iteration,

$$\underline{x}^{(k+1)} = \underline{x}^{(k)} + \underline{\delta x}^{(k)} \tag{5.13}$$

The algorithm for solving a system of nonlinear algebraic equations via the multivariate Newton-Raphson method follows analogously from the single variable version. The steps are as follows:
1. Make an initial guess for $\underline{x}$.
2. Calculate the Jacobian and the Residual at the current value of $\underline{x}$.
3. Solve equation (5.9) for $\underline{\delta x}^{(k)}$.
4. Calculate $\underline{x}^{(k+1)}$ from equation (5.13).
5. If the solution has not converged, loop back to step 2.

The multivariate Newton-Raphson Method suffers from the same short-comings as the single-variable Newton-Raphson Method. Specifically, as with all methods for solving nonlinear algebraic equations, you need a good initial guess. Second, the method does provide fast (quadratic) convergence until you are close to the solution. Third, if the determinant of the

Jacobian is zero, the method fails. This last constraint is the multi-dimensional analogue of the fact that the single variable Newton-Raphson method diverged when the derivative was zero.

The determination of convergence of a system that has multiple variables requires a tolerance. One could use a tolerance on each variable. That is the relative error on x_i must be less than tol_i. Alternatively, one can use something like the root mean square (RMS) error,

$$err_{RMS}^{(k)} = \sqrt{\frac{1}{n}\sum_{i=1}^{n}\left(\frac{x_i^{(k)} - x_i^{(k-1)}}{x_i^{(k-1)}}\right)^2} \tag{5.14}$$

to provide a single error for the entire system. In this case, even with an RMS relative error on x of 10^{-m}, you are not guaranteed that every variable has m good significant digits. You are only guaranteed that the RMS error is less than the acceptable tolerance.

Let's work two examples.

Example 5.1. Multivariate Newton-Raphson Method

Consider the system of two nonlinear algebraic equations.

$$f_1(x_1, x_2) = (x_1)^2 + (x_2)^2 - 4 = 0$$
$$f_2(x_1, x_2) = (x_1)^2 - (x_2) + 1 = 0$$

The first equation is that of a circle with radius 2 centered at the origin. The second equation is that of a parabola. In Figure 5.1. We plot the solution to the two equations independently. Since there are two variables in each equation, there are an infinite number of solutions for the equations treated independently. The solution to the system of nonlinear algebraic equations corresponds to ordered pairs of (x_1, x_2) that satisfy both equations. In Figure 5.1., this corresponds to the intersection between the two curves.

Any solution technique finds only one root at a time. From the plot we can estimate that one of the roots is near $(x_1, x_2) = (1, 2)$.

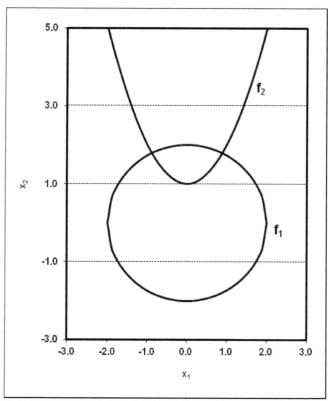

Figure 5.1. Plot of $f_j(x_1, x_2) = 0$ for $j = 1$ and 2.

In order to use the multivariate Newton-Raphson method, we must first determine the functional form of the n^2 partial derivatives analytically. For this small system, we have

$$J_{1,1} = \left(\frac{\partial f_1}{\partial x_1}\right) = 2x_1 \quad J_{1,2} = \left(\frac{\partial f_1}{\partial x_2}\right) = 2x_2$$

$$J_{2,1} = \left(\frac{\partial f_2}{\partial x_1}\right) = 2x_1 \quad J_{2,2} = \left(\frac{\partial f_2}{\partial x_2}\right) = -1$$

The residual is composed simply of the functions,

$$\underline{R} = \begin{bmatrix} f_1 \\ f_2 \end{bmatrix} = \begin{bmatrix} (x_1)^2 + (x_2)^2 - 4 \\ (x_1)^2 - (x_2) + 1 \end{bmatrix}$$

We now follow the algorithm outlined above.

Step One. Make an initial guess. $(x_1, x_2) = (1, 2)$

Step Two. Using that initial guess, calculate the residual and the Jacobian.

$$\underline{\underline{J}}^{(1)} = \begin{bmatrix} 2 & 4 \\ 2 & -1 \end{bmatrix} \quad \text{and} \quad \underline{R}^{(1)} = \begin{bmatrix} 1 \\ 0 \end{bmatrix}$$

Step Three. Solve equation (5.9) for $\underline{\delta x}^{(k)}$.
$$\underline{\delta x}^{(1)} = \begin{bmatrix} -0.1 \\ -0.2 \end{bmatrix}$$

Step 4. Calculate $\underline{x}^{(k+1)}$ from equation (5.13).
$$\underline{x}^{(2)} = \begin{bmatrix} 0.9 \\ 1.8 \end{bmatrix}$$

Step 5. If the solution has not converged, loop back to step 2.

Further iterations yield

	$\underline{x}$	$\underline{\underline{J}}$	$\underline{R}$	$\underline{\delta x}$
1	$\begin{bmatrix} 1 \\ 2 \end{bmatrix}$	$\begin{bmatrix} 2 & 4 \\ 2 & -1 \end{bmatrix}$	$\begin{bmatrix} 1 \\ 0 \end{bmatrix}$	$\begin{bmatrix} -0.1 \\ -0.2 \end{bmatrix}$
2	$\begin{bmatrix} 0.9 \\ 1.8 \end{bmatrix}$	$\begin{bmatrix} 1.8 & 3.6 \\ 1.8 & -1 \end{bmatrix}$	$\begin{bmatrix} 0.05 \\ 0.01 \end{bmatrix}$	$\begin{bmatrix} -0.0104 \\ -0.0087 \end{bmatrix}$

Systems of Nonlinear Algebraic Equations - 87

3	$\begin{bmatrix} 0.8896 \\ 1.7913 \end{bmatrix}$	$\begin{bmatrix} 1.7792 & 3.5826 \\ 1.7792 & -1.0000 \end{bmatrix}$	$\begin{bmatrix} 0.1835e-3 \\ 0.1079e-3 \end{bmatrix}$	$\begin{bmatrix} -0.6991e-4 \\ -0.1650e-4 \end{bmatrix}$
4	$\begin{bmatrix} 0.8895 \\ 1.7913 \end{bmatrix}$	$\begin{bmatrix} 1.7791 & 3.5826 \\ 1.7791 & -1.0000 \end{bmatrix}$	$\begin{bmatrix} 0.5159e-8 \\ 0.4887e-8 \end{bmatrix}$	$\begin{bmatrix} -0.2780e-8 \\ -0.0059e-8 \end{bmatrix}$

So one of the two roots is located at $(x_1, x_2) = (0.8895, 1.7913)$. (The other is located at $(x_1, x_2) = (-0.8895, 1.7913)$ by symmetry. From an examination of the final value of $\underline{\delta x}$ both solutions have converged to an absolute error on x of less than 10^{-8}.

Example 5.2. Multivariate Newton-Raphson Method: Linear Systems

Linear systems are a subset of non-linear systems. The multivariate Newton-Raphson solve linear systems exactly in one iteration, just as was the case in the single-variable problem.
Consider the system of linear equations:

$$f_1(x_1, x_2) = 5(x_1) + (x_2) - 4 = 0$$
$$f_2(x_1, x_2) = (x_1) - 3(x_2) + 1 = 0$$

The solutions of the two independent equations is plotted in Figure 5.2. The solution of the system of equations is the intersection of the two lines. For linear equations, the Jacobian is a constant matrix,

$$\underline{\underline{J}}^{(k)} = \begin{bmatrix} 5 & 1 \\ 1 & -3 \end{bmatrix}$$

The residual is composed simply of the functions,

$$\underline{R} = \begin{bmatrix} f_1 \\ f_2 \end{bmatrix} = \begin{bmatrix} 5x_1 + x_2 - 4 \\ x_1 - 3x_2 + 1 \end{bmatrix}$$

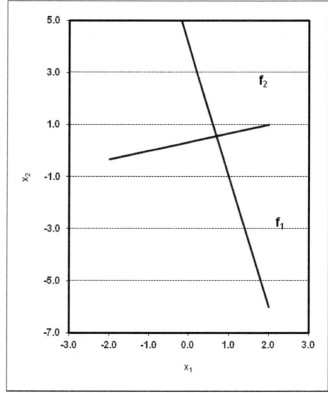

Figure 5.2. Plot of $f_j(x_1, x_2) = 0$ for $j = 1$ and 2.

We now follow the algorithm outlined above.

Step One. Make an initial guess.
$(x_1, x_2) = (2, 2)$
Step Two. Using that initial guess, calculate the residual and the Jacobian.

$$\underline{\underline{J}}^{(1)} = \begin{bmatrix} 5 & 1 \\ 1 & -3 \end{bmatrix} \quad \text{and} \quad \underline{R}^{(1)} = \begin{bmatrix} 8 \\ -3 \end{bmatrix}$$

Step Three. Solve equation (5.9) for $\underline{\delta x}^{(k)}$. $\quad \underline{\delta x}^{(1)} = \begin{bmatrix} -1.3125 \\ -1.4375 \end{bmatrix}$

Step 4. Calculate $\underline{x}^{(k+1)}$ from equation (5.13). $\quad \underline{x}^{(2)} = \begin{bmatrix} 0.6875 \\ 0.5625 \end{bmatrix}$

A second iteration will show that this is the exact solution.

5.3. Multivariate Newton-Raphson Method with Numerical Derivatives

There is a pretty obvious drawback to the Multivariate Newton-Raphson method, namely that one must provide the analytical form of *n*x*n* partial derivatives. If n is small, it can be sometimes be done. If *n* is large, this process is simply not practical. Therefore, just as was the case for the single variable Newton-Raphson method, we turn to numerical methods to provide estimates of the partial derivatives. The second-order centered-finite difference formula for the first derivative was derived in Chapter 3,

$$\left. \frac{df}{dx} \right|_{x_i} = \frac{f(x_{i+1}) - f(x_{i-1})}{2h} + O(h^2) \tag{3.7}$$

This can be directly extended to partial derivatives as follows,

$$\left(\left. \frac{df_j}{dx_i} \right|_{x_i} \right)_{x_{m \neq i}} = \frac{f_j(x_1, x_2, \ldots x_i + h_i, \ldots x_n) - f_j(x_1, x_2, \ldots x_i - h_i, \ldots x_n)}{2h_i} + O(h^2) \tag{5.15}$$

Of note, only the value of x_i changes when the differentiation is with respect to x_i. Furthermore, the value of discretization, h_i, may be different for each variable, x_i. Each iteration of a multivariate Newton-Raphson method with analytical derivatives requires *n* function evaluations and n^2 derivative evaluations. Each iteration of a multivariate Newton-Raphson method with numerical derivatives requires $n+2n^2$ function evaluations. So there is more computational effort in the numerical scheme, but potentially a much reduced effort in developing the code since only one function must be entered.

A MATLAB code which implements the multivariate Newton-Raphson method with numerical derivatives is provided later in this chapter.

5.4. Subroutine Codes

In this section, we provide a routine for implementing the multivariate Newton-Raphson method with numerical derivatives. Note that these codes correspond to the theory and notation exactly as laid out in this book. These codes do not contain extensive error checking, which would complicate the coding and defeat their purpose as learning tools. That said, these codes work and can be used to solve problems.

As before, on the course website, two entirely equivalent versions of this code are provided and are titled *code.m* and *code_short.m*. The short version is presented here. The longer version, containing instructions and serving more as a learning tool, is not presented here. The numerical mechanics of the two versions of the code are identical.

Code 5.1. Multivariate Newton-Raphson with Numerical derivatives (nrndn_short)

```
function [x,err,f] = nrndn(x0,tol,iprint)
%
% inputs:
% x0 = initial guess of x, column vector of length n
% tol = RMS tolerance on relative error on x, scalar
% iprint = value of 1 requests iteration information
% f(x) entered in the function "funkeval" at the bottom of this file
%
% outputs:
%
% x = converged solution, column vector of length n
% f = RMS value of f, scalar
% err = RMS value of relative error on x, scalar
%
maxit = 1000;
n = max(size(x0));
Residual = zeros(n,1);
Jacobian = zeros(n,n);
InvJ = zeros(n,n);
dx = zeros(n,1);
x = zeros(n,1);
xold = zeros(n,1);
xeval = zeros(n,1);
xp = zeros(2);
fp = zeros(n,n,2);
dxcon = zeros(n,1);
dxcon(1:n) = 0.01;
x = x0;
err = 100.0;
iter = 0;
while ( err > tol )
    for j = 1:1:n
```

```
         dx(j) = min(dxcon(j)*x(j),dxcon(j));
         i2dx(j) = 1.0/(2.0*dx(j));
      end
      Residual = funkeval(x);
      for j = 1:1:n
         for i = 1:1:n
            xeval(i) = x(i);
         end
         xp(1) = x(j) - dx(j);
         xp(2) = x(j) + dx(j);
         for k = 1:1:2
            xeval(j) = xp(k);
            fp(:,j,k) = funkeval(xeval);
         end
      end
      for i = 1:1:n
         for j = 1:1:n
            Jacobian(i,j) = i2dx(j)*( fp(i,j,2) - fp(i,j,1) );
         end
      end
      xold = x;
      invJ = inv(Jacobian);
      deltax = -invJ*Residual;
      for j = 1:1:n
         x(j) = xold(j) + deltax(j);
      end
      iter = iter +1;
      err = sqrt( sum(deltax.^2) /n  ) ;
      f = sqrt(sum(Residual.*Residual)/n);
      if (iprint == 1)
         fprintf (1,'iter = %4i, err = %9.2e f = %9.2e \n ', iter, err, f);
      end
      if ( iter > maxit)
         Residual
         error ('maximum number of iterations exceeded');
      end
end

function f = funkeval(x)
n = max(size(x));
f = zeros(n,1);
f(1) = x(1)^2 + x(2)^2 - 4;
f(2) = x(1)^2 - x(2) + 1;
```

An example of using nrndn_short is given below.

```
» [x,err,f] = NRNDN_short([2,2],1.0e-6,1)
iter =     1, err = 5.83e-001 f = 3.54e+000
 iter =    2, err = 1.91e-001 f = 6.60e-001
 iter =    3, err = 2.78e-002 f = 7.31e-002
 iter =    4, err = 6.13e-004 f = 1.54e-003
 iter =    5, err = 2.99e-007 f = 7.52e-007
```

```
x   =     0.8895      1.7913
err =     2.9893e-007
f   =     7.5211e-007
```

The root is at $(x_1, x_2) = (0.8895, 1.7913)$ and the error is less than the tolerance of 10^{-6}.

5.5. Problems

Problem 5.1.
Find the solution to the following system of nonlinear algebraic equations near (1,1,1).
$$f_1 = x_1 + 2x_2 + 3x_3 - 4$$
$$f_2 = x_1^3 - 4x_2^3$$
$$f_3 = x_3 - \sin x_3$$

Problem 5.2.
Perform 2 iterations of multivariate Newton-Raphson on the following system of nonlinear algebraic equations.

$$f_1 = x_1^2 + x_2^2 - 4$$
$$f_2 = x_1^2 - x_2 + 1$$

Show the value of the Jacobian, residual and $\delta x^{(k)}$ at each iteration. Use and initial guess of $(x_1, x_2) = (1, 2)$. Report the values of (x_1, x_2) for the first two iterations.

Problem 5.3.
An understanding of thermodynamics allow us to describe chemical equilibrium. If we put chemicals A and B in a pot and they react to form chemical C, we can determine how much of A, B and C are present under either (a) isothermal conditions where the pot is maintained at a constant temperature or (b) adiabatic conditions where the pot is insulated and no heat is allowed to escape. If we have three chemicals, we can consider that we have three unknowns, the mole fraction of each component, but it turns out those three unknowns can be related to a single variable, the extent of reaction, χ. The relationship between the number of moles and the extent of reaction is

$$n_{final} = n_{initial} + v\chi$$

where we have to know the initial number of moles in the pot and the stoichiometric cofficient, v. If we want the final mole fractions, we can relate them to the final moles via:

$$x_i = \frac{n_i}{\sum_{j=1}^{3} n_j}$$

In the isothermal case, the only equation is an expression for the chemical equilibrium between reactants and products. This expression is coded up in a MatLab input file for use with nrnd.m and provided below.

In the adiabatic case, we have an additional variable, the unknown adiabatic temperature, which is determined through an energy balance, which is also provided in the input file below.

Consider the reaction:

½ N_2 + 3/2 H_2 ←→ NH_3

Consider a batch reactor (a closed pot) with initially 0.5 moles of N_2, 1.5 moles of H_2, and 0.0 moles of NH_3, all initially at 25 C.

(a) Solve for the extent of reaction if the reactor is operated isothermally at T_{iso} = 520.0 K and the pressure is 1 atm. (Set the variable adiabatic = 0 for this case.) Also use the equations provided above to report final mole fractions of each component.

(b) Solve for the extent of reaction and adiabatic temperature if the reactor is operated adiabatically K and the pressure is 111 atm. (Set the variable adiabatic = 1 for this case.) Also use the equations provided above to report final mole fractions of each component.

To solve this problem, you require a bunch of physical constants. These are already entered in the input file, but are provided here again for the sake of completeness.

Heats of formation (Kcal/mol)
```
%       nitrogen    hydrogen    ammonia
Hf  =  [0;          0;          -10.960];
```

Free energies of formation (Kcal/mol)
```
%       nitrogen    hydrogen    ammonia
Gf  =  [0.0;        0.0;        -3.903];
```

Heat capacity constants (a in first row, b in second row, c in third row, d in fourth row)
```
%               nitrogen    hydrogen    ammonia
Cpcon  =       [6.903       6.952       6.5846
                -0.03753    -0.04576    0.61251
                0.1930      0.09563     0.23663
                -0.6861     -0.2079     -1.5981];
```

Systems of Nonlinear Algebraic Equations - 93

Input file for Problem 5.3:

```
function f = funkeval(x)
%
%  these two lines force a column vector of length n
%
n = max(size(x));
f = zeros(n,1);
%
%  enter the functions here
%
%
%  adiabatic chemical reaction equilibria calculation
%  nitrogen plus hydrogen to ammonia
%  (Sandler Illustration 9.1-4 page 509)
%
%
%  Checklist.
%       (1)  Make sure you have entered the correct number of reactions (nr)
%       (2)  Make sure you have entered the correct number of components (nc)
%       (3)  Set adiabatic = 1 for adiabatic system, set adiabatic = 0 for isothermal system
%       (4)  If the system is isothermal, set the isothermal temperature , (Tiso)
%       (5)  Enter initial mole amounts of each component (flowin)
%       (6)  Enter initial temperature of each component (Tin)
%       (7)  Enter reactor pressure in atmospheres (pressure)
%       (8)  Enter the stoichiometric matrix
%       (9)  Enter the heats of formation  (Kcal/mol)
%       (10) Enter the free energies of formation (Kcal/mol)
%       (11) Enter the heat capacity constants
%
% STEP ONE.  ENTER PROBLEM SPECIFICATIONS
%
%    (1)  number of independent chemical reactions
%
nr = 1;
%
%    (2) number of components
%
nc = 3;
%
%    (3)  determines whether the system is adiabatic or isothermal
%
adiabatic = 0;
%adiabatic = 1;
%
%    (4)  isothermal temperature
%
Tiso = 520;
%
%    (5)  inlet flowrates
%
flowin=zeros(nc,1);
flowin(1) = 0.5;
flowin(2) = 1.5;
%
%    (6) inlet temperatures (K)
%
Tin=zeros(nc,1);
TempC = 25;
Tin(1) = TempC+273.1;
Tin(2) = TempC+273.1;
%
%    (7)  reactor pressure in atmospheres
%
pressure = 1;   % put this in atmospheres
%
%    (8)  stoichiometric matrix
%    the order is
%    N2 H2 NH3
%
nu=[-1/2    -3/2    1];
%
%    (9) heats of formation at the reference temperature
%    from Sandler (Kcal/mol)
%    N2 H2 NH3
%
Hf = [0;    0;   -10.960];
%
%    (10)  free energies of formation at the reference temperature
%    from Sandler (Kcal/mol)
%    N2 H2 NH3
%
Gf = [0.0;  0.0;    -3.903];
%
%    (11)  heat capacity constants from Sandler Appendix
```

```matlab
% Each column contains parameters for a given component.
% The four rows contain the four parameters a, b, c, d for
% Cp = 4.184*(a + b*T/10^2 + c*T^2/10^5 +d*T^3/10^9) [Joules/mole/K]
% N2 H2 NH3
%
Cpcon = [6.903       6.952        6.5846
        -0.03753    -0.04576       0.61251
         0.1930     0.09563        0.23663
        -0.6861    -0.2079        -1.5981];
%
%     you won't need to touch anything below here.
%
%
% STEP TWO:   LET THE PROGRAM DO ITS WORK
%
%
% STEP TWO A:   Identify unknowns
%
%
% nr extents of reaction
%
X=zeros(nr,1);
for i = 1:nr
   X(i) = x(i);
end
%
% 1 Temperature
%
if (adiabatic == 1)
   T = x(nr+1);
else
   T=Tiso;
end
%
% STEP TWO B.   DEFINE KNOWNS
%
Hf = Hf*4.184*1000; % (Joules/mol)
Gf = Gf*4.184*1000; % (Joules/mol)
%
%  reference temperature
%
Tref = 25 + 273.1;
%
%  constants
%
R = 8.314; % [J/mol/k]
RT = R*T;
%
%
%
% heats of reaction at the reference temperature
%
DHrref = nu*Hf;
%
% free energies of reaction at at the reference temperature
%
DGrref = nu*Gf;
%
% scale heat capacity constants
%
for k = 1:nc
   Cpcon(2,k) = Cpcon(2,k)*10^-2;
   Cpcon(3,k) = Cpcon(3,k)*10^-5;
   Cpcon(4,k) = Cpcon(4,k)*10^-9;
end
Cpcon = 4.184*Cpcon;
%Cp = 4.184*(Cpcon(1,2) + Cpcon(2,2)*T/10^2 + Cpcon(3,2)*T^2/10^5 +Cpcon(4,2)*T^3/10^9)
%
% enthalpies at arbitrary T
% (need this for flow out of the reactor term in energy balance)
%
for k = 1:nc
   Cpint2(k) = Cpint(T,Cpcon,k) - Cpint(Tref,Cpcon,k);
end
%
% heats of reaction at arbitrary T
% (need this for energy balance)
%
for i = 1:nr
   DHr(i) = DHrref(i);
   for k = 1:nc
      DHr(i) = DHr(i) + nu(i,k)*Cpint2(k);
   end
end
%
% calculate equilibrium constants at the reference temperature
%
for i = 1:nr
```

```matlab
        Karef(i) = exp(-DGrref(i)/(R*Tref));
end
%
% integrate the heat of reaction over RT^2 from Tref to T
%  (need this to calculate Ka as a function of Temperature)
%
for i = 1:nr
    term1 = DHroRT2intfunk(T,Tref,Cpcon,nu,DHrref,R,i,nc);
    term2 = DHroRT2intfunk(Tref,Tref,Cpcon,nu,DHrref,R,i,nc);
    DHroRT2int2(i) = term1 - term2;
end
%
% calculate equilibrium constants at arbitrary  temperature
%
for i = 1:nr
    Ka(i) = Karef(i)*exp(DHroRT2int2(i));
end
%
% define molar composition based on extent of reactions
%
xE = flowin;
for i = 1:nr
    for k = 1:nc
        xE(k) = xE(k) + nu(i,k)*X(i);
    end
end
E = sum(xE);
xE = xE/E;
%
% stream enthalpies
%
heatin = zeros(nc,1);
for k = 1:nc
    heatin(k) = flowin(k)*(Cpint(Tin(k),Cpcon,k) - Cpint(Tref,Cpcon,k));
end
heatintot = sum(heatin);
heatout = zeros(nc,1);
for k = 1:nc
    heatout(k) = E*xE(k)*(Cpint(T,Cpcon,k) - Cpint(Tref,Cpcon,k));
end
heatouttot = sum(heatout);
%
% Step Three.  Write down nr+1 equations
%
%
% nr equilibrium contraints
%
for i = 1:nr
    f(i) = 1;
    for k = 1:nc
        f(i) = f(i)*(xE(k)^nu(i,k));
    end
    f(i) = f(i) - (pressure^(-sum(nu(i,:))))*Ka(i);
end
%
% nc material balances
%
%fmole =  flowin + nu'*X - E*xE;
%for k = 1:nc
%    f(k+nr) = fmole(k);
%end
%
% 1 mole fraction constraint
%
%f(nr+nc+1) = sum(xE) - 1;
%
% 1 energy balance
%
convaid = 0.01;
heatrxn = 0.0;
for i = 1:nr
    heatrxn = heatrxn + DHr(i)*X(i);
end
if (adiabatic == 1)
    f(nr+1) = convaid*(heatintot - heatouttot - heatrxn);
end

%%%fprintf (1,'xE = %15e  %15e  %15e \n', xE(1), xE(2), xE(3))

%
% integrated heat capacity of component k
%
function y = Cpint(T,Cpcon,k)
%Cp = 4.184*(a + b*T/10^2 + c*T^2/10^5 +d*T^3/10^9) [Joules/mole/K]
y = Cpcon(1,k)*T + Cpcon(2,k)*T^2/2 + Cpcon(3,k)*T^3/3 + Cpcon(4,k)*T^4/4;
%[Joules/mole]
%
```

```
% integrate the heat of reaction over RT^2 from Tref to T
%
function y = DHroRT2intfunk(T,Tref,Cpcon,nu,DHrref,R,i,nc)
term1a = -DHrref(i)/(R*T);
term1b = 0.0;
for k = 1:nc
    t1b1 = Cpcon(1,k)*log(T) + Cpcon(2,k)/2*T + Cpcon(3,k)/6*T^2 + Cpcon(4,k)/12*T^3;
    t1b2 = Cpcon(1,k)*Tref + Cpcon(2,k)*Tref^2/2 + Cpcon(3,k)*Tref^3/3 + Cpcon(4,k)*Tref^4/4;
    term1b = term1b + nu(i,k)/R*(t1b1 + t1b2/T);
end
y = term1a + term1b;
```

Chapter 6. Numerical Integration

6.1. Introduction

Many undergraduates prefer differentiation to integration because the rules for differentiation are few and once they are known virtually any analytical function can be readily differentiated. Integration, although it is the inverse operation of differentiation, is much less loved because, depending upon one's familiarity, analytical integration involves searching through integral tables with no hope that an analytical form of the integral even exists.

Thus numerical integration is welcomed because it provides a simple and methodical procedure to evaluate integrals, without resorting to tables and regardless of the existence of an analytical form. That said, if an analytical form of the integral exists, it is preferable to use it. From a physical point of view, having an analytical function gives us insight into the parameters appearing within the function. From a professional point of view, we will be laughed at by our peers if we try to present work using numerical integration where an analytical form was readily apparent. Still, if there is a problem to be solved and no analytical integral in sight, numerical integration can, more often than not, come to the rescue.

6.2. Trapezoidal Rule

The Trapezoidal rule gets its name from the use of trapezoids to approximate integrals. Consider that you want to integrate a function, $f(x)$, from a to b. The trapezoidal rule says that the integral of that function can be approximated by a trapezoidal with a base of length $(b-a)$ and sides of height $f(a)$ and $f(b)$. Graphically, the trapezoidal rule is represented in Figure 6.1.

The single-interval trapezoidal rule is expressed as

$$\int_a^b f(x)dx \approx \frac{1}{2}(f(a)+f(b))(b-a)$$

(6.1)

The right hand side of equation (6.1) is the expression for the area of a trapezoid, shown in the Figure 6.1. Now, it is quite easy to imagine a case where the single-interval trapezoidal rule is going to give a terrible estimate. Consider the curve shown in Figure 6.2. (top). In order to increase the accuracy of the trapezoidal rule, one can approximate the integral by many smaller trapezoids. One can see in Figure 6.2. (bottom) that as the number of trapezoids increases, the area of the integral not accounted for by the trapezoidal rule decreases.

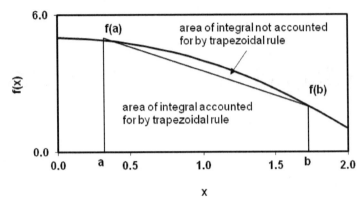

Figure 6.1. Schematic of trapezoidal rule for integration.

If we are integrating from a to b using n trapezoids (or intervals), then the base of each trapezoid (or discretization in the x-dimension) is

$$h = \frac{b-a}{n} \quad (6.2)$$

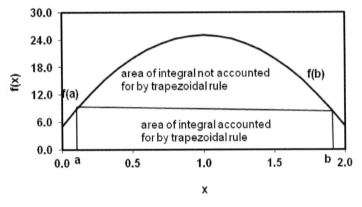

The area of each of these smaller trapezoids, A_i, is

$$A_i = \frac{h}{2}\left(f(x_i) + f(x_{i+h})\right) \quad (6.3)$$

where the position of each point, x_i, is given

$$x_i = a + (i-1)*h \quad (6.4)$$

for $i = 1$ to $n+1$. We note that if there are n intervals, there must be n+1 points, with $x_1 = a$ and $x_{n+1} = b$.

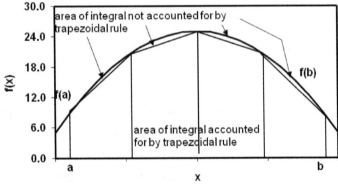

Figure 6.2. Accuracy of the trapezoidal rule improves as the number of trapezoids used increases.

The integral is given by the summation of the areas of all the trapezoids:

$$\int_a^b f(x)dx \approx \frac{h}{2}\sum_{i=1}^{n}\left(f(x_i)+f(x_{i+h})\right) \tag{6.5}$$

The quantity $f(x_i)$ appears twice in the summation of equation (6.5). It appears once as the left-hand-side of the trapezoid that forms the ith interval and it occurs once as the right-hand-side of the trapezoid that forms the i-1th interval. This is true of all $f(x_i)$ except the endpoints, $f(a)$ and $f(b)$. For these reasons, equation (6.5) can be algebraically manipulated to yield:

$$\int_a^b f(x)dx \approx \frac{h}{2}\left[f(a)+f(b)+2\sum_{i=2}^{n}f(x_i)\right] \tag{6.6}$$

This is the most common form of the mutliple-interval trapezoidal rule. The accuracy of the trapezoidal rule increases as *n* increases. The trapezoidal rule is a first order method, which means two things. First, a first-order polynomial (a straight line) has been used to approximate the along each interval. Second, the error is proportional to the interval size to the first power. Halving the size of the interval halves the error. We postpone a comparative discussion of the accuracy of various methods until more methods have been introduced. Suffice it to say that the trapezoidal rule is not very accurate and we would prefer when possible to use higher order methods.

A MATLAB code which implements the trapezoidal rule is provided later in this chapter.

6.2. Second-Order Simpson's Rule

The Second-Order Simpson's Rule (often called the Simpson's 1/3 Rule although not in this book) is another technique used for numerical integration. A more accurate approach to integration involves the use of higher order polynomial approximating the integrand. The Second-Order Simpson's Rules uses a second-order (quadratic) polynomial.

The application of the 2nd order Simpson's rule is demonstrated graphically in Figure 6.3. The

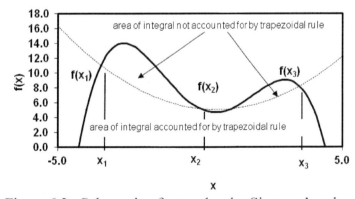

Figure 6.3. Schematic of second-order Simpson's rule for integration.

function is evaluated at three points (or equivalently across two intervals). A parabola is fit through those three points. Since there is a simple analytical formula for the integral of the parabola, the integration is then done analytically.

The derivation of the second-order Simpson's rule for numerical integration is an interesting application of linear algebra and regression skills that we already know from Chapters 1 and 2. Consider that we have three equally spaced points, x_1, x_2 and x_3. The x_2 and x_3 are related to x_1 by $x_2 = x_1 + \Delta x$ and $x_3 = x_1 + 2\Delta x$. We also know the function value evaluated at these three points, $f(x_1)$, $f(x_2)$, $f(x_3)$.

We are going to fit a second-order polynomial (parabola) to these three points. Since a parabola has three coefficients, we can perfectly fit the three points, $(x_1, f(x_1))$, $(x_2, f(x_2))$ and $(x_3, f(x_3))$. Substituting these three points into a second-order polynomial yields

$$c_2 x_1^2 + c_1 x_1 + c_0 - f(x_1) = 0$$
$$c_2 x_2^2 + c_1 x_2 + c_0 - f(x_2) = 0$$
$$c_2 x_3^2 + c_1 x_3 + c_0 - f(x_3) = 0$$

We have a system of three algebraic equations. Moreover, the equations are linear in the unknown coefficients. We can write these equations in matrix form, $\underline{\underline{A}}\,\underline{x} = \underline{b}$, where

$$\underline{\underline{A}} = \begin{bmatrix} x_1^2 & x_1 & 1 \\ x_2^2 & x_2 & 1 \\ x_3^2 & x_3 & 1 \end{bmatrix}, \quad \underline{x} = \begin{bmatrix} c_2 \\ c_1 \\ c_0 \end{bmatrix}, \quad \underline{b} = \begin{bmatrix} f(x_1) \\ f(x_2) \\ f(x_3) \end{bmatrix}$$

The solution to this system of equations is

$$\underline{x} = \begin{bmatrix} c_2 \\ c_1 \\ c_0 \end{bmatrix} = \begin{bmatrix} \dfrac{f(x_1) - 2f(x_2) + f(x_3)}{2\Delta x^2} \\ -\dfrac{f(x_1)[2x_1 + 3\Delta x] - f(x_2)[4x_1 + 4\Delta x] + f(x_3)[2x_1 + \Delta x]}{2\Delta x^2} \\ \dfrac{f(x_1)[x_1^2 + 3x_1 \Delta x + 2\Delta x^2] - f(x_2)[2x_1^2 + 4x_1 \Delta x] + f(x_3)[x_1^2 + x_1 \Delta x]}{2\Delta x^2} \end{bmatrix}$$

At this point we have the coefficients of our parabola. Now we want to integrate the parabola from x_1 to x_3.

$$I_{Simp} = \int_{x_1}^{x_3} \left(c_2 x^2 + c_1 x + c_0 \right) dx = \left[\frac{c_2 x^3}{3} + \frac{c_1 x^2}{2} + c_0 x \right]_{x_1}^{x_3}$$

$$I_{Simp} = \left[\frac{c_2 x_3^3}{3} + \frac{c_1 x_3^2}{2} + c_0 x_3\right] - \left[\frac{c_2 x_1^3}{3} + \frac{c_1 x_1^2}{2} + c_0 x_1\right]$$

We substitute in the formulae for c_2, c_1, and c_0. We also substitute in $x_2 = x_1 + \Delta x$ and $x_3 = x_1 + 2\Delta x$. After a lot of messy but straightforward algebra, we find that the expression simplifies to:

$$I_{Simp} = \frac{\Delta x}{3}\left[f(x_1) + 4f(x_2) + f(x_3)\right] \tag{6.7}$$

This is the second-order Simpson's rule where we have only 2 intervals (one parabola). If we divide the interval up into many intervals, then we have to sum each of the integrated intervals up. Just as was the case with the trapezoidal rule, the first and last points appear only once. Each interior point at the beginning or end of a parabola will be counted twice. Each point in the center of the parabola is counted once, but has a weighting factor of four. So, our final second-order Simpson's rule for *n*-intervals (*n*+1 points) is:

$$I_{Simp} = \frac{\Delta x}{3}\left[f(x_1) + 4\sum_{\substack{i=2 \\ \text{even}}}^{n} f(x_2) + 2\sum_{\substack{i=3 \\ \text{odd}}}^{n-1} f(x_2) + f(x_{n+1})\right] \tag{6.8}$$

Again, as was the case for the Trapezoidal rule, the second-order Simpson's rule provides greater accuracy if the range is broken up into a number of intervals. For the the second-order Simpson's rule, the number of intervals, n, must be even because, as you can see in the above plot, each polynomial fit requires 2 intervals.

For the same number of intervals, the second-order Simpson's rule is more accurate than the trapezoidal rule because we used a high-order polynomial to fit the original function. Because the second-order Simpson's rule is second order, the error is proportional to the interval size to the second power. Halving the size of the interval reduces the error by a factor of four. We postpone a comparative discussion of the accuracy of various methods until more methods have been introduced.

A MATLAB code which implements the second-order Simpson's rule is provided later in this chapter.

6.3. Higher Order Simpson's Rules

The third-order Simpson's rule uses a third order (cubic) polynomial fit over three intervals to approximate the integral. The derivation of the method is precisely analogous to that for the second-order method. We present the result with proof. For a single application (three intervals or four points), the integral is

$$I_{Simp} = \frac{3\Delta x}{8}[f(x_1) + 3f(x_2) + 3f(x_3) + f(x_4)] \qquad (6.9)$$

If we want to use many intervals, then we need to add up these individual components. We must also use a number of intervals that is a multiple of 3.

$$I_{Simp} = \frac{3\Delta x}{8}\left[f(x_1) + 3\sum_{i=2,5,8}^{n-1} f(x_2) + 3\sum_{i=3,6,9}^{n} f(x_2) + 2\sum_{i=4,7,10}^{n-2} f(x_2) + f(x_{n+1})\right] \qquad (6.10)$$

The fourth-order Simpson's rule uses a fourth order (quartic) polynomial fit over four intervals to approximate the integral. For a single application (four intervals or five points), the integral is

$$I_{Simp} = \frac{2\Delta x}{45}[7f(x_1) + 32f(x_2) + 12f(x_3) + 32f(x_4) + 7f(x_5)] \qquad (6.11)$$

If we want to use many intervals, then we need to add up these individual components. We must also use a number of intervals that is a multiple of 4.

$$I_{Simp} = \frac{2\Delta x}{45}\left[7f(x_1) + 32\sum_{i=2,6,10}^{n-2} f(x_2) + 12\sum_{i=3,7,11}^{n-1} f(x_2) + 32\sum_{i=4,8,12}^{n} f(x_2) + 14\sum_{i=5,9,13}^{n-3} f(x_2) + 7f(x_{n+1})\right]$$
$$(6.12)$$

MATLAB codes which implement the third-order and fourth-order Simpson's rule are provided later in this chapter.

6.5. Quadrature

Quadrature takes a slightly different approach to the numerical evaluation of integrals. Quadrature is based on the assumption that we can get a better estimate of the integral with fewer function evaluations if we use non-equally spaced points at which to evaluate the function. The determination of these points and the weighting coefficients that correspond to each data point follows a methodical procedure. We do not derive them here.

The integral for the nth order Gaussian Quadrature is given by:

$$I_{quad} = \sum_{i=1}^{n} c_i f(x_i) \qquad (6.13)$$

The particular values of the weighting constants, c, and the points where we evaluate the function, x, can be taken from a table in any numerical methods text book.

The advantage of a quadrature technique is that it can be quite accurate for very few function evaluations. Thus it is much faster and if need to repeatedly evaluate integrals it is the method of choice. For the evaluation of a couple integrals, the Simpson's Rules are better because we know we can increase accuracy by increasing the number of intervals used.

MATLAB codes which implement Gaussian Quadrature for 2nd to 6th order are provided later in this chapter.

6.6. Example

We want to evaluate the change in entropy of methane for an isothermal expansion or compression. To describe the thermodynamic state of the fluid, we will rely on tried and true friend, the van der Waals equation of state.

$$p = \frac{RT}{V-b} - \frac{a}{V^2} \qquad (4.16)$$

where P is pressure (Pa), T is temperature (K), V is molar volume (m³/mol), R is the gas constant (8.314 J/mol/K = 8.314 Pa*m³/mol/K), a is the van der Waal's attraction constant (.2303 Pa*m⁶/mol² for methane) and b is the van der Waal's repulsion constant (4.306x10⁻⁵ m³/mol for methane). The change in entropy for an isothermal expansion without phase change is

$$\Delta S = \int_{V_1}^{V_2} \left(\frac{\partial P}{\partial T}\right)_{V'} dV'$$

The partial derivative of the pressure with respect to temperature at constant molar volume for a van der Waal's gas is obtained from differentiating the van der Waals EOS.

$$\left(\frac{\partial P}{\partial T}\right)_V = \frac{R}{V-b}$$

so the change in entropy is

$$\Delta S = \int_{V_1}^{V_2} \frac{R}{V-b} dV'$$

We can analytically evaluate this integral so as to provide a basis of comparison for the accuracy of the various numerical techniques. The analytical integral is .

$$\Delta S = R \ln\left(\frac{V_2 - b}{V_1 - b}\right)$$

The pressure as a function of molar volume is shown in Figure 6.4. for van der Waals methane at 298 K. The partial derivative, $\left(\frac{\partial P}{\partial T}\right)_V$, as a function of molar volume is also shown in Figure 6.4. for van der Waals methane at 298 K. The deriviative is the function we will need to numerically integrate. in order to obtain the entropy difference. It is a smooth, monotonically decreasing function in our range of interest.

Let's expand the gas from 0.03 m³/mol to 0.1 m³/mol. We compare results using the analytical solution, and several numerical methods in Table 6.1. What is immediately obvious is that as we increase the number of intervals for a given order of method, we see a gradual increase in accuracy. In quadrature as we increase the order of the method, we decrease the error by an order of magnitude. Also, as we increase the order of a Simpson's-type method, we see a drastic increase in accuracy. For example it takes 100,000 intervals in the Trapezoidal rule to equal the accuracy of 100 intervals in the Simpson's Fourth Order Method. Quadrature is even more efficient. A 6-point quadrature equals the accuracy of 1000 trapezoidal intervals in this example.

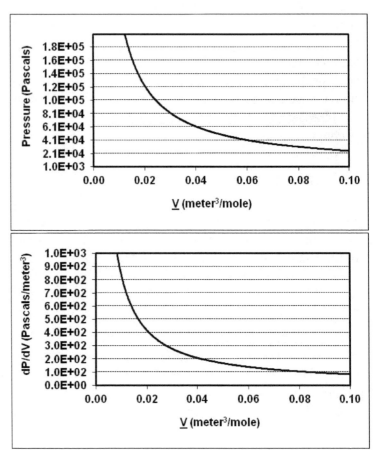

Figure 6.4. (top) Pressure as a function of molar volume at 298 K for van der Waals methane. (bottom) $\left(\frac{\partial P}{\partial T}\right)_V$ as a function of molar volume at 298 K for van der Waals methane.

Technique	# of intervals	$\Delta S(J/mol/K)$	percent error
analytical	-	10.0182	0.0
Trapezoidal	1	12.62476	2.60E+01
Trapezoidal	2	10.79212	7.73E+00
Trapezoidal	3	10.38034	3.61E+00
Trapezoidal	4	10.22639	2.08E+00
Trapezoidal	10	10.05242	3.42E-01
Trapezoidal	100	10.01854	3.44E-03
Trapezoidal	1000	10.01819	3.44E-05
Trapezoidal	10000	10.01819	3.44E-07
Trapezoidal	100000	10.01819	3.44E-09
Simpson's 2nd Order	2	10.18124	1.63E+00
Simpson's 2nd Order	4	10.03781	1.96E-01
Simpson's 2nd Order	10	10.01892	7.30E-03
Simpson's 2nd Order	100	10.01819	8.17E-07
Simpson's 2nd Order	1000	10.01819	8.18E-11
Simpson's 3rd Order	3	10.09979	8.14E-01
Simpson's 3rd Order	6	10.02738	9.18E-02
Simpson's 3rd Order	9	10.02040	2.21E-02
Simpson's 3rd Order	99	10.01819	1.91E-06
Simpson's 3rd Order	999	10.01819	1.85E-10
Simpson's 4th Order	4	10.02825	1.00E-01
Simpson's 4th Order	8	10.01869	4.96E-03
Simpson's 4th Order	100	10.01819	3.39E-09
Simpson's 4th Order	1000	10.01819	3.55E-14
Quadrature 2nd Order	2	9.91943	9.86E-01
Quadrature 3rd Order	3	10.00942	8.75E-02
Quadrature 4th Order	4	10.01743	7.63E-03
Quadrature 5th Order	5	10.01812	6.61E-04
Quadrature 6th Order	6	10.01819	5.70E-05
quad (MATLAB)	?	10.01828	9.20e-04
quad8 (MATLAB)	?	10.01819	3.36e-08

Table 6.1. Comparison of accuracy of various numerical integration methods.

6.7. Multidimensional Integrals

Integrals over an area or a volume are not uncommon in science and engineering. The most conceptually straight forward approach would be to extend the existing one-dimensional Simpson's and quadrature methods to multidimensional integrals through sequential application. Let us examine this application, for the simplest case, which involves the Trapezoidal rule for two-dimensional integration with fixed limits of integration. This essentially involves integrating a function over a rectangular area.

A two-dimensional integral with fixed limits of integration can be written as

$$I_{2D} = \int_{xo}^{xf}\int_{yo}^{yf} f(x,y)dydx = \int_{xo}^{xf} g(x)dx \tag{6.14}$$

where the integrand of the outermost integral, $g(x)$, is

$$g(x) = \int_{yo}^{yf} f(x,y)dy \tag{6.15}$$

This integrand has no dependence on y since that functionality has been integrated out. As a reminder, the 1-D trapezoidal rule using n intervals ($n+1$ points) is

$$\int_a^b f(x)dx \approx \frac{h}{2}\left[f(a) + f(b) + 2\sum_{i=2}^{n} f(x_i)\right] \tag{6.6}$$

We apply the trapezoidal rule to the integral over y only first and substitute that into $g(x)$

$$g(x) = \int_{yo}^{yf} f(x,y)dy \approx \frac{h_y}{2}\left[f(x,y_o) + f(x,y_f) + 2\sum_{i=2}^{n_y} f(x,y_i)\right] \tag{6.16}$$

Substituting the discretized approximation for $g(x)$ in equation (6.16) into equation (6.14) yields

$$\int_{xo}^{xf}\int_{yo}^{yf} f(x,y)dydx \approx \int_{xo}^{xf} \frac{h_y}{2}\left[f(x,y_o) + f(x,y_f) + 2\sum_{i=2}^{n_y} f(x,y_i)\right]dx \tag{6.17}$$

Well, we can repeat the application of the trapezoidal rule:

$$I_{2D} \approx \frac{h_x}{2} \left\{ \begin{array}{l} \frac{h_y}{2}\left[f(x_o,y_o)+f(x_o,y_f)+2\sum_{i=2}^{n_y}f(x_o,y_i)\right] \\ +\frac{h_y}{2}\left[f(x_f,y_o)+f(x_f,y_f)+2\sum_{i=2}^{n_y}f(x_f,y_i)\right] \\ +2\sum_{j=2}^{n_x}\frac{h_y}{2}\left[f(x_j,y_o)+f(x_j,y_f)+2\sum_{i=2}^{n_y}f(x_j,y_i)\right] \end{array} \right\} \quad (6.18)$$

Now we can simplify this as much as possible,

$$I_{2D} \approx \frac{h_x h_y}{4}\left\{ \begin{array}{l} f(x_o,y_o)+f(x_o,y_f)+f(x_f,y_o)+f(x_f,y_f)+4\sum_{i=2}^{n_y}\sum_{j=2}^{n_x}f(x_j,y_i) \\ +2\sum_{i=2}^{n_y}\left[f(x_o,y_i)+f(x_f,y_i)\right]+2\sum_{j=2}^{n_x}\left[f(x_j,y_o)+f(x_j,y_f)\right] \end{array} \right\} \quad (6.19)$$

If we add up the number of function evaluations, we can see that we have $(n_x+1)(n_y+1)$ function evaluations. If $n_x = n_y = n$, then we have $(n+1)^2$ function evaluations for a 2-D integral. By extension, if we need to evaluate an *m*-dimensional integral, then we will have $(n+1)^m$ function evaluations. In other words, the number of function evaluations scales exponentially with the dimensionality of the system. By observation we observe that there are four points with a weighting factor of one corresponding to the four corners of the area. There are four sums with a weighting factor of two corresponding to the four edges of the area. There is one double sum with a weighting factor of four corresponding to the interior of the area.

This simple procedure which culminates in the 2-D Trapezoidal rule can be applied to higher dimensions. For example, the 3-D Trapezoidal rule applied to an integral with fixed limits of integration yields a corresponding formula. This volume corresponds to a right parallelepiped. By observation we observe that there are eight points with a weighting factor of one corresponding to the eight vertices of the volume. There are twelve sums with a weighting factor of two corresponding to the twelve edges of the volume. There are six double sums with a weighting factor of four corresponding to the six faces of the volume. There is one triple sum with a weighting factor of eight corresponding to the interior of the volume.

$$I_{3D} = \frac{h_x h_y h_z}{2^3} \begin{bmatrix} \begin{Bmatrix} f(x_o,y_o,z_o) + f(x_o,y_o,z_f) + f(x_o,y_f,z_o) + f(x_o,y_f,z_f) \\ + f(x_f,y_o,z_o) + f(x_f,y_o,z_f) + f(x_f,y_f,z_o) + f(x_f,y_f,z_f) \end{Bmatrix} \\ + 2 \begin{Bmatrix} \sum_{i=2}^{n_z} [f(x_o,y_o,z_i) + f(x_o,y_f,z_i) + f(x_f,y_o,z_i) + f(x_f,y_f,z_i)] \\ + \sum_{j=2}^{n_y} [f(x_o,y_j,z_o) + f(x_o,y_j,z_f) + f(x_f,y_j,z_o) + f(x_f,y_j,z_f)] \\ + \sum_{k=2}^{nx} [f(x_k,y_o,z_o) + f(x_k,y_o,z_f) + f(x_k,y_f,z_o) + f(x_k,y_f,z_f)] \end{Bmatrix} \\ + 4 \begin{Bmatrix} \sum_{j=1}^{n_y} \sum_{i=2}^{n_z} [f(x_o,y_j,z_i) + f(x_f,y_j,z_i)] \\ + \sum_{k=2}^{nx} \sum_{i=2}^{n_z} [f(x_k,y_o,z_i) + f(x_k,y_f,z_i)] \\ + \sum_{k=2}^{nx} \sum_{j=1}^{n_y} [f(x_k,y_j,z_o) + f(x_k,y_j,z_f)] \end{Bmatrix} \\ + 8 \sum_{k=2}^{nx} \sum_{j=1}^{n_y} \sum_{i=2}^{n_z} f(x_k,y_j,z_i) \end{bmatrix} \quad (6.20)$$

This is the explicit form of the trapezoidal rule applied in 3-dimensions, when the limits of integration are constant

This simple procedure which culminates in the multidimensional Trapezoidal rule can be applied to higher order methods. For example, the two-dimensional Simpon's second order method applied to an integral with constant limits of integration can be derived in an analogous manner and is presented in equation (6.20). The simplified final expression involves terms with weighting factors of one (the corners of the area), weighting factors of two (odd-numbered nodes on the edges of the area), weighting factors of four (even-numbered nodes on the edges of the area and odd-odd double sums in the interior), weighting factors of eight (odd-even double sums in the interior) and weighting factors of sixteen (even-even double sums in the interior).

For any of these methods, if the boundary of the area over which the integration occurs is not regular, simple functions like that given in equation (6.19), (6.20) and (6.21) cannot be written. However, the same concept can still be applied. Integration over y can be performed, generating a one-dimensional function $g(x)$, which can then be integrated.

For high dimensional integrals with complicated geometries, completely different methods of numerical integration are often invoked, including Monte Carlo integration, which randomly samples the function. Such methods are beyond the scope of this introductory text.

$$I_{2D} \approx \frac{h_x h_y}{9} \begin{Bmatrix} f(x_o,y_o) + f(x_o,y_f) + f(x_f,y_o) + f(x_f,y_f) \\ + 2\left(\sum_{i=3,5,7}^{n_y-2}[f(x_o,y_i) + f(x_f,y_i)] + \sum_{j=3,5,7}^{n_x-2}[f(x_j,y_o) + f(x_j,y_f)] \right) \\ + 4\left(\sum_{i=2,4,6}^{n_y-1}[f(x_o,y_i) + f(x_f,y_i)] + \sum_{j=2,4,6}^{n_x-1}[f(x_j,y_o) + f(x_j,y_f)] \right) \\ + 8\left(\sum_{j=2,4,6}^{n_x-1}\sum_{i=3,5,7}^{n_y-2}f(x_j,y_i) + \sum_{j=3,5,7}^{n_x-2}\sum_{i=2,4,6}^{n_y-1}f(x_j,y_i) \right) \\ + 4\sum_{j=3,5,7}^{n_x-2}\sum_{i=3,5,7}^{n_y-2}f(x_j,y_i) + 16\sum_{j=2,4,6}^{n_x-1}\sum_{i=2,4,6}^{n_y-1}f(x_j,y_i) \end{Bmatrix} \quad (6.21)$$

6.8. Subroutine Codes

In this section, we provide a routine for implementing the various numerical integration methods described above. Note that these codes correspond to the theory and notation exactly as laid out in this book. These codes do not contain extensive error checking, which would complicate the coding and defeat their purpose as learning tools. That said, these codes work and can be used to solve problems.

As before, on the course website, two entirely equivalent versions of this code are provided and are titled *code.m* and *code_short.m*. The short version is presented here. The longer version, containing instructions and serving more as a learning tool, is not presented here. The numerical mechanics of the two versions of the code are identical.

Code 6.1. Trapezoidal Rule (trapezoidal_short)

```
function integral = trapezoidal_short(a,b,nintervals);
dx = (b-a)/nintervals;
npoints = nintervals + 1;
x_vec = [a:dx:b];
integral = funkeval(x_vec(1));
for i = 2:1:nintervals
   integral = integral + 2*funkeval(x_vec(i));
end
integral = integral + funkeval(x_vec(npoints));
integral = 0.5*dx*integral;
fprintf(1,'\nUsing the Trapezoidal method \n');
fprintf(1,'to integrate from %f to %f with %i nintervals,\n',a,b,nintervals);
fprintf(1,'the integral is %e \n \n',integral);
```

```
function f = funkeval(x)
f = x^2 + 5 - sin(x);
```

An example of using trapezoidal_short is given below.

```
» integral = trapezoidal_short(0.03,0.1,100);

Using the Trapezoidal method
to integrate from 0.030000 to 0.100000 with 100 nintervals,
the integral is 3.457785e-001
```

Code 6.2. Simpson's Second Order Rule (simpson2_short)

```
function integral = simpson2_short(a,b,nintervals);
if (mod(nintervals,2) ~= 0)
   fprintf('Simpsons 2nd method requires an even number of intervals.\n');
else
   dx = (b-a)/nintervals;
   npoints = nintervals + 1;
   x_vec = [a:dx:b];
   integral_first = funkeval(x_vec(1));
   integral_last = funkeval(x_vec(npoints));
   integral_4 = 0.0;
   for i = 2:2:nintervals
         integral_4 = integral_4 + funkeval(x_vec(i));
   end
   integral_2 = 0.0;
   for i = 3:2:nintervals-1
         integral_2 = integral_2 + funkeval(x_vec(i));
   end
   integral = integral_first + integral_last + 4.0*integral_4 + 2.0*integral_2;
   integral = dx/3*integral;
   fprintf(1,'\nUsing the Simpsons Second Order method \n');
   fprintf(1,'to integrate from %f to %f with %i nintervals,\n',
a,b,nintervals);
   fprintf(1,'the integral is %e \n \n',integral);
end

function f = funkeval(x)
f = x^2 + 5 - sin(x);
```

An example of using simpson2_short is given below.

```
» integral = simpson2_short(0.0,1.0,100)

Using the Simpsons Second Order method
to integrate from 0.000000 to 1.000000 with 100 nintervals,
the integral is 4.873636e+000
```

Integration - 111

Code 6.3. Simpson's Third Order Rule (simpson3_short)

```
function integral = simpson3_short(a,b,nintervals);
if (mod(nintervals,3) ~= 0)
      fprintf('Simpsons 3rd Order method requires a # of intervals that is a
multiple of 3.\n');
else
   dx = (b-a)/nintervals;
   npoints = nintervals + 1;
   x_vec = [a:dx:b];
   integral_first = funkeval(x_vec(1));
   integral_last = funkeval(x_vec(npoints));
   integral_3a = 0.0;
   for i = 2:3:nintervals-1
        integral_3a = integral_3a + funkeval(x_vec(i));
   end
   integral_3b = 0.0;
   for i = 3:3:nintervals
        integral_3b = integral_3b + funkeval(x_vec(i));
   end
   integral_2 = 0.0;
   for i = 4:3:nintervals-2
        integral_2 = integral_2 + funkeval(x_vec(i));
   end
   integral = integral_first + integral_last + 3.0*integral_3a ...
       + 3.0*integral_3b + 2.0*integral_2;
   integral = 3.0*dx/8.0*integral;
   fprintf(1,'\nUsing the Simpsons Third Order method \n');
   fprintf(1,'to integrate from %f to %f with %i
nintervals,\n',a,b,nintervals);
   fprintf(1,'the integral is %e \n \n',integral);
end

function f = funkeval(x)
f = x^2 + 5 - sin(x);
```

An example of using simpson3_short is given below.

```
» integral = simpson3_short(0.0,1.0,99)

Using the Simpsons Third Order method
to integrate from 0.000000 to 1.000000 with 99 nintervals,
the integral is 4.873636e+000
```

Code 6.4. Simpson's Fourth Order Rule (simpson4_short)

```
function integral = simpson4_short(a,b,nintervals);
if (mod(nintervals,4) ~= 0)
      fprintf('Simpsons 4th Order method requires a # of intervals that is a
multiple of 4.\n');
else
```

```
   dx = (b-a)/nintervals;
   npoints = nintervals + 1;
   x_vec = [a:dx:b];
   integral_first = funkeval(x_vec(1));
   integral_last = funkeval(x_vec(npoints));
   integral_32a = 0.0;
   for i = 2:4:nintervals-2
        integral_32a = integral_32a + funkeval(x_vec(i));
   end
   integral_32b = 0.0;
   for i = 4:4:nintervals
        integral_32b = integral_32b + funkeval(x_vec(i));
   end
   integral_12 = 0.0;
   for i = 3:4:nintervals-1
        integral_12 = integral_12 + funkeval(x_vec(i));
   end
   integral_14 = 0.0;
   for i = 5:4:nintervals-3
        integral_14 = integral_14 + funkeval(x_vec(i));
   end
   integral = 7.0*integral_first + 7.0*integral_last + 32.0*integral_32a ...
      + 32.0*integral_32b + 12.0*integral_12 + 14.0*integral_14;
   integral = 2.0*dx/45.0*integral;
   fprintf(1,'\nUsing the Simpsons Fourth Order method \n');
   fprintf(1,'to integrate from %f to %f with %i
nintervals,\n',a,b,nintervals);
   fprintf(1,'the integral is %e \n \n',integral);
end

function f = funkeval(x)
f = x^2 + 5 - sin(x);
```

An example of using simpson4_short is given below.

```
» integral = simpson4_short(0.0,1.0,100)

Using the Simpsons Fourth Order method
to integrate from 0.000000 to 1.000000 with 100 nintervals,
the integral is 4.873636e+000
```

Code 6.5. Gaussian Quadrature (gaussquad_short)

```
function integral = gaussquad_short(a,b,norder);
if (norder < 2 | norder > 6)
   fprintf('This code only works for order between 2 and 6\n');
else
   a0 = 0.5*(b+a);
   a1 = 0.5*(b-a);
   if (norder == 2)
      c(1) = 1.0;
```

```
      c(2) = c(1);
      x_table(1) = -0.577350269;
      x_table(2) = -x_table(1);
   elseif (norder == 3)
      c(1) = 0.555555556;
      c(2) = 0.888888889;
      c(3) = c(1);
      x_table(1) = -0.774596669;
      x_table(2) = 0.0;
      x_table(3) = -x_table(1);
   elseif (norder == 4)
      c(1) = 0.347854845;
      c(2) = 0.652145155;
      c(3) = c(2);
      c(4) = c(1);
      x_table(1) = -0.861136312;
      x_table(2) = -0.339981044;
      x_table(3) = -x_table(2);
      x_table(4) = -x_table(1);
   elseif (norder == 5)
      c(1) = 0.236926885;
      c(2) = 0.478628670;
      c(3) = 0.568888889;
      c(4) = c(2);
      c(5) = c(1);
      x_table(1) = -0.906179846;
      x_table(2) = -0.538469310;
      x_table(3) = 0.0;
      x_table(4) = -x_table(2);
      x_table(5) = -x_table(1);
   elseif (norder == 6)
      c(1) = 0.171324492;
      c(2) = 0.360761573;
      c(3) = 0.467913935;
      c(4) = c(3);
      c(5) = c(2);
      c(6) = c(1);
      x_table(1) = -0.932469514;
      x_table(2) = -0.661209386;
      x_table(3) = -0.238619186;
      x_table(4) = -x_table(3);
      x_table(5) = -x_table(2);
      x_table(6) = -x_table(1);
end
integral = 0.0;
for i = 1:1:norder
   x(i) = a0 + a1*x_table(i);
   f(i) = funkeval(x(i));
   integral = integral + c(i)*f(i);
end
integral = integral*a1;
fprintf(1,'\nUsing %i order Gaussian Quadrature \n', norder);
fprintf(1,'to integrate from %f to %f \n',a,b);
```

```
    fprintf(1,'the integral is %e \n \n',integral);
end

function f = funkeval(x)
R = 8.314;
b = 4.306e-5;
f = R/(x-b);
```

An example of using guassquad_short is given below.

```
» integral = gaussquad_short(0.03,0.1,4)

Using 4 order Gaussian Quadrature
to integrate from 0.030000 to 0.100000
the integral is 1.001743e+001
```

6.9. Problems

Problem 6.1.
Consider the normal distribution

$$f(x;\mu,\sigma) = \frac{1}{\sqrt{2\pi}\sigma} e^{-\frac{1}{2}\left(\frac{x-\mu}{\sigma}\right)^2}$$

This function does not have an analytical integral.
For the standard normal distribution, where the mean is zero and the standard deviation is one, evaluate the integral from x = -2.0 to 1.0, i.e. $p(-2.0 \le x \le 1.0)$, using

(a) the trapezoidal method with 1 interval.
(b) the trapezoidal method with 10 intervals.
(c) the trapezoidal method with 100 intervals.
(d) the trapezoidal method with 1000 intervals.
(e) the Simpson's Second Order method with 100 intervals.
(f) the Simpson's Second Order method with 1000 intervals.
(g) the Simpson's Third Order method with 99 intervals.
(h) the Simpson's Fourth Order method with 100 intervals.
(i) Gaussian quadrature of sixth order.
(j) the cdf command in MatLab.
(k) Comment on the effect of number of intervals and order of the method.

Problem 6.2..

Consider the van der Waals equation of state.

$$P = \frac{RT}{\underline{V}-b} - \frac{a}{\underline{V}^2}$$

where P is pressure (Pa), T is temperature (K), $\underline{V}$ is molar volume (m³/mol), R is the gas constant (8.314 J/mol/K = 8.314 Pa*m³/mol/K), a is the van der Waal's attraction constant (0.2303 Pa*m⁶/mol² for methane) and b is the van der Waal's repulsion constant (4.306x10⁻⁵ m³/mol for methane).

The entropy change upon expanding is

$$\Delta S = \int_{\underline{V}_1}^{\underline{V}_2} \left(\frac{\partial P}{\partial T}\right)_{\underline{V}'} d\underline{V}'$$

where

$$\left(\frac{\partial P}{\partial T}\right)_{\underline{V}} = \frac{R}{\underline{V}-b}$$

Find the entropy change upon expanding methane from 0.05 to 0.11 m³/mol at T = 298 K.

(a) Find analytical integral.
(b) Find the integral using Gaussian quadrature.
(c) Find the integral using the data provided in 'file.hw11p02c.txt'.
(d) Repeat part (a) at T = 398 K. Comment.

Chapter 7. Solution of Ordinary Differential Equations

7.1. Introduction

The dynamic behavior of many relevant systems and materials can be described with ordinary differential equations (ODEs). In this chapter, we provide an introduction to the techniques for numerical solution of ODEs. We begin with a single, first-order ODE initial value problem. We then extend the process to high-order ODEs, systems of ODEs and boundary value problems. The techniques described in this chapter work for both linear and nonlinear ODEs. However, we point out that for linear ODEs and some non-linear ODEs there may exist more elegant analytical solutions. In this book, we do not investigate the analytical solutions but limit ourselves to numerical methods that provide valid solutions.

7.2. Initial Value Problems

Regardless of whether one intends to solve an ordinary differential equation (ODE) with analytical or numerical techniques, the problem must first be properly posed. Let us initially limit ourselves to a single, first-order ODE, involving a single independent variable x and a single dependent variable $y(x)$. Differential equations that involve more than one independent variable are called partial differential equations (PDEs) and are not considered in this book. Differential equations that involve more than one dependent variable constitute systems of ODEs and addressed later in this chapter. The solution to an ODE is a function, $y(x)$. The most general formulation of this ODE is

$$f(x, y(x), y'(x)) = 0 \qquad (7.1)$$

where we have invoked the shorthand notation, $y'(x) \equiv dy/dx$, for the derivative of y with respect to x. Often, it is possible through algebraic manipulation of equation (7.1) to isolate the derivative on the LHS,

$$y'(x) = f(x, y(x)) \qquad (7.2)$$

This is perhaps the most familiar form of an ODE, but it is not the most general.

ODEs are frequently categorized as linear and nonlinear ODEs. It is important to remember that the linearity of the ODE is defined by the linearity of y only (and specifically not of x). The unknown y must be operated on exclusively by linear operators in order for the ODE to be considered linear. (Recall that the differential operator is a linear operator.) The general form of a linear first order ODE is

$$a(x)y'(x) + b(x)y(x) + c(x) = 0 \tag{7.3}$$

The coefficients, $a(x)$, $b(x)$ and $c(x)$ can be nonlinear functions of x.

Regardless of whether one is solving an ODE of the form given in (7.1), (7.2) or (7.3), the problem is not properly posed and does not have a unique solution until an initial condition is supplied. The initial condition provides a value of the function at one point.

$$y(x = x_0) = y_0 \tag{7.4}$$

The initial value problem (IVP) is the combination of the ODE and the initial condition.

7.3. Euler Method

It is instructive to begin our study of numerical techniques for solving ODEs with a first order method. In practice, we will never use a first-order method to solve an ODE, due to its low accuracy. Nevertheless, the first-order method plays an important role as an instructional tool.

The Euler method relies on a Taylor series truncated after the linear term.

$$f(x_{i+1}) = f(x_i) + \frac{df}{dx}\bigg|_{x_i} h + O(h^2) \tag{3.2}$$

which can be rewritten in the nomenclature of ODEs as

$$y(x_{i+1}) = y(x_i) + \Delta x \, y'(x_i) \tag{7.5}$$

where the discretization is given by $\Delta x = x_{i+1} - x_i$. Equation (7.5) is the Euler method. One begins at the initial condition $(x, y) = (x_0, y_0)$ as given in the problem statement. One choose a discretization, Δx, which is a crucial decision. One then evaluates the derivative at x_0 using the ODE (equation (7.1), (7.2) or (7.3)), which also was given in the problem statement. The Euler method in equation (7.5) can then be used directly to estimate the value of the unknown function at

$x_1 = x_0 + \Delta x$. This process can be repeated moving further down the x-axis for as long as interest remains.

Before we introduce specific examples, two points of discussion are necessary. First, the choice of discretization or step-size is crucial. It is true that the smaller the step-size the more accurate the solution. However, we must balance the desire for accuracy with the need for computational efficiency. If our choice of step-size is so small that we need an intractable number of steps, then that doesn't help us. If the step size is too big, the method will fail. In fact, for the simple methods presented in this book, **the only reason a numerical routine for solving an ODE will fail is because the step-size was too large**. The reason can be seen in Figure 7.1. In the top figure, it is clear that as the step-size increases, the error increases. The error is quantified in the middle figure. For some functions it is possible that the error leads to values of the function that are not permitted, such as negative values of $y(x)$ in functions that require a square root.

The second point of interest is the determination of when to stop the Euler iterations. The final value of x is simply chosen by the user. Frequently, the stopping point is determined when one has reached a steady state. For example in the bottom part of Figure 7.1, the function clearly goes to zero. Continuing to solve the ODE out to large values of x typically serves no useful purpose. If one doesn't know how long it will take to reach a steady state, then one may have to guess the final value. If it is too short, then one can simply extend the Euler procedure using the final point of the previous

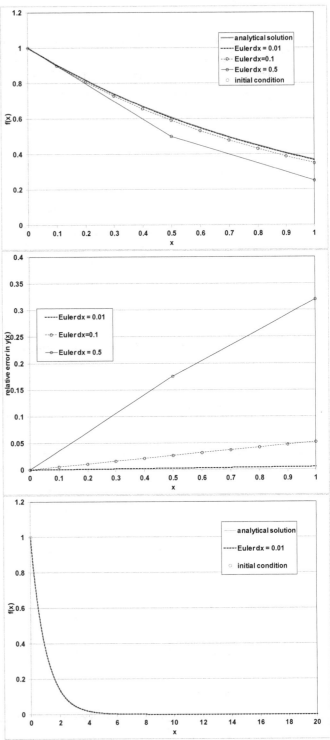

Figure 7.1 Application of Euler method to $y'(x) = -y$ with initial condition $y(x=0)=1$. Top: solution to x=1. Middle: error up to x=1. Bottom: To x=20.

process as the initial condition for the continuation.

A MATLAB code which implements the Euler method for a single first-order ODE is provided later in this chapter.

7.4. Classical Fourth-Order Runge-Kutta Method

The obvious solution to inaccuracy of the Euler method is to increase the order of the method. There are numerous flavors of ODE solvers of all orders. Here we simply present the Classical Fourth-Order Runge-Kutta (RK4) Method, which is a tried and true solution technique. The RK4 method proceeds as does the Euler method by starting at the initial condition and following an estimate of the average slope over the interval, $\langle y' \rangle$, out to the next discretization point.

$$y(x_{i+1}) = y(x_i) + \Delta x \langle y' \rangle \tag{7.6}$$

In the Euler method, the estimate of the average slope was $\langle y' \rangle = y'(x_i)$, simply the value of the slope at the beginning of the interval. The RK4 method uses a higher order approximation for the average slope over the interval, namely

$$\langle y' \rangle = \frac{1}{6}(k_1 + 2k_2 + 2k_3 + k_4) \tag{7.7}$$

where

$$\begin{aligned}
k_1 &= y'(x_i, y(x_i)) \\
k_2 &= y'\left(x_i + \frac{\Delta x}{2}, y(x_i) + \frac{\Delta x}{2}k_1\right) \\
k_3 &= y'\left(x_i + \frac{\Delta x}{2}, y(x_i) + \frac{\Delta x}{2}k_2\right) \\
k_4 &= y'(x_i + \Delta x, y(x_i) + \Delta x k_3)
\end{aligned} \tag{7.8}$$

Without a rigorous derivation, we can see that the Runge-Kutta method improves the quality of the estimate of the slope over the interval by evaluating the slope at the beginning, middle (twice) and end of the interval. The RK4 method is a fourth-order method, so the error decreases as the step-size to the fourth power.

In Figure 7.2, we present an application of the RK4 method. In the left figure we show that even for coarse discretization, the RK4 method can be very accurate. In the right figure we quantify the relative error. Note that the error is now shown on a log axis

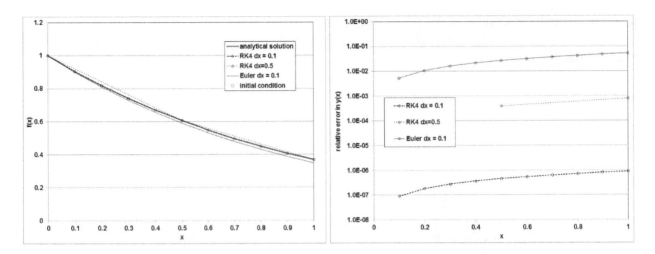

Figure 7.2. Application of RK4 method to $y'(x)=-y$ with initial condition $y(x=0)=1$. Left: solution up to x=1. Right: error up to x=1.

A MATLAB code which implements the classical fourth-order Runge-Kutta method for a single first-order ODE is provided later in this chapter.

7.5. Application to Systems of Ordinary Differential Equations

The extension of the Euler or Runge-Kutta method to systems of ODEs is very straightforward for an initial value problem. Let's suppose that we have n ODEs, each of which can be written in the form of equation (7.2) as

$$\frac{dy_1(x)}{dx} = f_1(x, y_1, y_2, y_3, \ldots y_{n-1}, y_n)$$
$$\frac{dy_2(x)}{dx} = f_2(x, y_1, y_2, y_3, \ldots y_{n-1}, y_n) \quad (7.9)$$
$$\ldots$$
$$\frac{dy_n(x)}{dx} = f_n(x, y_1, y_2, y_3, \ldots y_{n-1}, y_n)$$

with n initial conditions:

$$y_j(x = x_0) = y_{j,0} \quad \text{for } j = 1 \text{ to } n \quad (7.10)$$

The fact that all of the conditions are given at the same value of x is what makes this problem an IVP. Note carefully, that there are n ODE's, thus n functions, f_j, and n unknown functions, y_j,

Ordinary Differential Equations - 121

but there is only one independent variable x. This is what makes this system a system of ODEs rather than PDEs. The techniques in this chapter will not solve PDEs.

The Euler method for a system of of ODEs can be written as

$$y_j(x_{i+1}) = y_j(x_i) + \Delta x * f_j(x_i, y_1(x_i), y_2(x_i), y_3(x_i), \ldots y_{n-1}(x_i), y_n(x_i)) \quad \text{for } j = 1 \text{ to } n \quad (7.11)$$

There is no difference between this equation and the equation for the single system ODE using Euler's method (equation (7.5)). In this case, remember that the subscript j attached to the y and the f denotes different functions. The subscript i attached to the x variable denotes steps (or iterations). The Euler algorithm for a system of ODEs simply requires one to evaluate all f_j at iteration i from the ODEs in the problem statement (equation 7.9), then compute all y_j at iteration $i+1$ from equation (7.11) and repeat the process.

The extension of RK4 to systems of ODE's is just as simple as the extension of Euler's method. However, because RK4 has a little more sophistication, the extension looks more complicated, when it is really not. The RK4 equation for a system of equations is given by

$$y_j(x_{i+1}) = y_j(x_i) + \left[\frac{1}{6}\left(k_{1,j} + 2k_{2,j} + 2k_{3,j} + k_{4,j}\right)\right]h \quad \text{for } j = 1 \text{ to } n \quad (7.12)$$

where

$$k_{1,j} = f_j(x_i, \{y_m(x_i)\})$$
$$k_{2,j} = f_j\left(x_i + \frac{\Delta x}{2}, \left\{y_m(x_i) + \frac{\Delta x}{2} k_{1,m}\right\}\right)$$
$$k_{3,j} = f_j\left(x_i + \frac{\Delta x}{2}, \left\{y_m(x_i) + \frac{\Delta x}{2} k_{2,m}\right\}\right) \quad \text{for } j = 1 \text{ to } n \quad (7.13)$$
$$k_{4,j} = f_j(x_i + \Delta x, \{y_m(x_i) + \Delta x k_{3,m}\})$$

In equation (7.13), the braces in $\{y_m(x_i)\}$ represent the set of all y_m from $m = 1$ to n., all of which are evaluated at x_i.

The multi-ODE RK4 algorithm requires that one first evaluate all $k_{1,j}$. With $k_{1,j}$ in hand, one can compute the arguments required for $k_{2,j}$. Once all of the $k_{2,j}$ are known, one can next compute the arguments required for $k_{3,j}$. Once all of the $k_{3,j}$ are known, one can next compute the arguments required for $k_{4,j}$. Once all the $k_{1,j}$, $k_{2,j}$, $k_{3,j}$ and $k_{4,j}$ are known, one can use the RK4 method (equation 7.12) to evaluate the value of the unknown functions at the next step, $y_j(x_{i+1})$.

MATLAB codes which implement the Euler method and the classical fourth-order Runge-Kutta method for a system of first-order ODEs are provided later in this chapter.

7.6. Higher-Order ODEs

Regardless of whether one is pursuing an analytical or numerical solution to a higher than first order ODE, there is a simple trick to converting an nth-order ODE to a system of n first-order ODEs. You make a substitution that transforms the nth-order differential equation into n first-order differential equations. Consider an nth-order ODE of the form

$$\frac{d^n y(x)}{dx^n} = f\left(x, y(x), \frac{dy(x)}{dx}, \frac{d^2 y(x)}{dx^2} \cdots \frac{d^{n-1} y(x)}{dx^{n-1}}\right) \tag{7.14}$$

with the following n initial conditions:

$$y(x = x_0) = y_0, \left.\frac{dy(x)}{dx}\right|_{x=x_0} = y'_0, \left.\frac{d^2 y(x)}{dx^2}\right|_{x=x_0} = y''_0, \ldots \left.\frac{d^{n-1} y(x)}{dx^{n-1}}\right|_{x=x_0} = y_0^{(n-1)} \tag{7.15}$$

The conversion is a three step process. The first step is defining n new variables, which always have the following form:

$$\begin{aligned} y_1(x) &= y(x) \\ y_2(x) &= \frac{dy(x)}{dx} \\ y_3(x) &= \frac{d^2 y(x)}{dx^2} \\ &\vdots \\ y_n(x) &= \frac{d^{n-1} y(x)}{dx^{n-1}} \end{aligned} \tag{7.16}$$

The second step is writing one first-order ODE for each of these n variables. The first n-1 ODEs have a trivial form. The last ODE is simply the original higher-order ODE, equation (7.14), with the new variables, equation (7.16) substituted in.

$$\frac{dy_1(x)}{dx} = y_2(x)$$

$$\frac{dy_2(x)}{dx} = y_3(x)$$

$$\vdots \qquad (7.16)$$

$$\frac{dy_{n-1}(x)}{dx} = y_n(x)$$

$$\frac{dy_n(x)}{dx} = f(x, y_1(x), y_2(x), y_3(x) \ldots y_{n-1}(x))$$

The third and final step of the conversion process is to rewrite the initial conditions, equation (7.15), in terms of the new variables,

$$y_1(x_0) = y_0, \; y_2(x_0) = y'_0, \; y_3(x_0) = y''_0, \ldots y_n(x_0) = y_0^{(n-1)} \qquad (7.17)$$

After the conversion equations (7.16) and (7.17) have constitute a system of n first-order ODEs with n initial conditions. This IVP can be solved using the methods of the previous section.

7.7. Boundary Value Problems

When one invokes Fick's law of diffusion in a material balance, or Fourier's law of heat conduction in an energy balance, or Newton's law of viscosity in a momentum balance, one can end up with a steady-state model that is a second order ODE, where the independent variable is position. What distinguishes these problems from other higher-order ODEs is that the conditions are not given at the same point. From a mathematical point of view, we can write this equation as

$$\frac{d^2 y(x)}{dx^2} = f\left(x, y(x), \frac{dy(x)}{dx}\right) \qquad (7.18)$$

with the following 2 boundary conditions:

$$y(x_0) = y_0 \qquad \text{and} \qquad y(x_f) = y_f \qquad (7.19)$$

Instead of being given the value of the function and its derivative at a single point, x_0, we now have a constraint on the function at two points, x_0 and x_f. For example, we know the temperature at either side of a metal rod, where our thermocouples are attached, but we know nothing about the derivative of the temperature. The boundary conditions in equation (7.19) define what is called the Boundary Value Problem (BVP).

As presented, the Euler or RK4 only work for IVPs, not BVPs. The solution lies in creating a numerical method that combines our ability to numerically solve nonlinear algebraic equations with ODEs. If we had been given an initial value of the first derivative, $\left.\frac{dy(x)}{dx}\right|_{x=x_0} = y'_0$, we would have an IVP and we could solve the problem easily, using the techniques in Section 7.6 (breaking up the second-order ODE into two first-order ODEs) and Section 7.5 (solving a system of first-order ODEs in an IVP). Therefore, we play to our strengths. We will guess a value of y'_0. We will solve the system of ODEs from x_0 to x_f. We will compare the calculated value of $y(x_f)$ with the boundary condition, y_f. If they match within a tolerance, we made a good guess. Probably, they don't match and we must make a new guess for y'_0 and iterate again. What should guide our guess for y'_0? Why, I have a just the thing! We should use the Newton-Raphson method with numerical derivatives for solving a single nonlinear algebraic equation. The unknown is y'_0. The algebraic equation is

$$g(y'_0) = y(x_f) - y_f = 0 \tag{7.20}$$

In other words, we must choose the correct initial condition on the derivative to satisfy the given boundary condition on the function. We understand that since $y(x_f)$ is arrived at via the solution of the set of ODEs, that changing the value of y'_0 will change the value of $y(x_f)$. Every evaluation of the function given in equation (7.20) requires a solution of the system of ODEs.

A MATLAB code which implements the solution of a BVP for a system of 2 ODEs using the Newton-Raphson method and the classical fourth-order Runge-Kutta method is provided later in this chapter.

7.8. Subroutine Codes

In this section, we provide routines for implementing the various numerical ODE solver methods described above. Note that these codes correspond to the theory and notation exactly as laid out in this book. These codes do not contain extensive error checking, which would complicate the coding and defeat their purpose as learning tools. That said, these codes work and can be used to solve problems.

As before, on the course website, two entirely equivalent versions of this code are provided and are titled *code.m* and *code_short.m*. The short version is presented here. The longer version, containing instructions and serving more as a learning tool, is not presented here. The numerical mechanics of the two versions of the code are identical.

Code 7.1. Euler Method – 1 ODE (euler1_short)

```
function [x,y]=euler1_short(n,xo,xf,yo);
dx = (xf-xo)/n;
x = zeros(n+1,1);
for i = 1:1:n+1
   x(i) = xo + (i-1)*dx;
end
y = zeros(n+1,1);
y(1) = yo;
for i =  1:1:n
   dydx = funkeval(x(i),y(i));
      y(i+1) = y(i) + dx*dydx;
end
close all;
iplot = 1;
if (iplot == 1)
   plot (x,y,'k-o'), xlabel( 'x' ), ylabel ( 'y' );
end
fid = fopen('euler1_out.txt','w');
fprintf(fid,'x                    y \n');
fprintf(fid,'%23.15e %23.15e    \n', [x,y]');
fclose(fid);

function dydx = funkeval(x,y);
dydx = -1.0*y;
```

An example of using euler1_short is given below.

» [x,y]=euler1_short(10,0,2,1);

This program generates outputs in three forms. First, the x and y vectors are stored in memory and can be directly accessed. Second, the program generates a plot of y vs. x. Third, the program generates an output file, *euler1_out.txt*, that contains x and y vectors in tabulated form.

Code 7.2. Fourth-Order Runge-Kutta Method – 1 ODE (rk41_short)

```
function [x,y]=rk41_short(n,xo,xf,yo);
dx = (xf-xo)/n;
x = zeros(n+1,1);
for i = 1:1:n+1
   x(i) = xo + (i-1)*dx;
end
y = zeros(n+1,1);
y(1) = yo;
for i =  1:1:n
   x1 = x(i);
   y1 = y(i);
   k1 = funkeval(x1,y1);
   x2 = x(i) + 0.5*dx;
```

```
        y2 = y(i) + 0.5*dx*k1;
        k2 = funkeval(x2,y2);
        x3 = x(i) + 0.5*dx;
        y3 = y(i) + 0.5*dx*k2;
        k3 = funkeval(x3,y3);
        x4 = x(i) + dx;
        y4 = y(i) + dx*k3;
        k4 = funkeval(x4,y4);
        dydx = 1.0/6.0*(k1 + 2.0*k2 + 2.0*k3 + k4);
        y(i+1) = y(i) + dx*dydx;
end
close all;
iplot = 1;
if (iplot == 1)
    plot (x,y,'k-o'), xlabel( 'x' ), ylabel ( 'y' );
end
fid = fopen('rk41_out.txt','w');
fprintf(fid,'x                y \n');
fprintf(fid,'%23.15e %23.15e    \n', [x,y]');
fclose(fid);

function dydx = funkeval(x,y);
dydx = -1.0*y
```

An example of using rk41_short is given below.

» [x,y]=rk41_short(10,0,2,1);

This command generates outputs in three forms. First, the x and y vectors are stored in memory and can be directly accessed. Second, the program generates a plot of y vs. x. Third, the program generates an output file, *rk41_out.txt*, that contains x and y vectors in tabulated form.

Code 7.3. Euler Method – n ODEs (eulern_short)

```
function [x,y]=eulern(n,xo,xf,yo);
dx = (xf-xo)/n;
x = zeros(n+1,1);
for i = 1:1:n+1
   x(i) = xo + (i-1)*dx;
end
m=max(size(yo));
y = zeros(n+1,m);
y(1,1:m) = yo(1:m);
dydx = zeros(m,1);
for i =   1:1:n
    dydx = funkeval(x(i),y(i,1:m));
       y(i+1,1:m) = y(i,1:m) + dx*dydx(1:m);
end
close all;
iplot = 1;
if (iplot == 1)
```

```
      for i = 1:1:m
         color_index = get_plot_color(i);
         plot (x(:),y(:,i),color_index);
         hold on;
      end
      hold off;
      xlabel( 'x' );
      ylabel ( 'y' );
      legend (int2str([1:m]'));
end
fid = fopen('eulern_out.txt','w');
fprintf(fid,'x   y(1) ... y(m) \n');
for i = 1:1:n+1
   fprintf(fid,'%23.15e ', x(i));
   for j = 1:1:m
      fprintf(fid,'%23.15e ', y(i,j));
   end
   fprintf(fid,' \n');
end
fclose(fid);

function dydx = funkeval(x,y);
dydx(1) = -1.0*y(1) - 2.0*y(2) - 0.5*y(3);
dydx(2) = -0.1*y(1) - 4.0*y(2) - 0.5*y(3);
dydx(3) = -0.5*y(1) - 0.4*y(2) - 0.2*y(3);
```

An example of using eulern_short is given below.

```
» [x,y]=eulern(100,0,10,[1,0,2]);
```

Note that the fourth input argument, yo, is now a vector. This program generates outputs in three forms. First, the x vector and y matrix are stored in memory and can be directly accessed. Second, the program generates a plot of y vs. x. Third, the program generates an output file, *eulern_out.txt*, that contains x and y in tabulated form.

Also note that this code calls an ancillary function for plotting, *get_plot_color*, which is reproduced below. The function assigns a different color to each function so that the curves are distinguishable on the graph. This function can be included at the bottom of the file eulern_short.m.

```
function color_index = get_plot_color(i);
if (i == 1)
   color_index = 'k-';
elseif (i == 2)
   color_index = 'r-';
elseif (i == 3)
   color_index = 'b-';
elseif (i == 4)
   color_index = 'g-';
elseif (i == 5)
```

```
      color_index = 'm-';
elseif (i == 6)
      color_index = 'k:';
elseif (i == 7)
      color_index = 'r:';
elseif (i == 8)
      color_index = 'b:';
elseif (i == 9)
      color_index = 'g:';
elseif (i == 10)
      color_index = 'm:';
else
      color_index = 'k-';
end
```

Code 7.4. Classical Fourth-Order Runge-Kutta Method – n ODEs (rk4n_short)

```
function [x,y]=rk4n_short(n,xo,xf,yo);
dx = (xf-xo)/n;
x = zeros(n+1,1);
for i = 1:1:n+1
   x(i) = xo + (i-1)*dx;
end
m=max(size(yo));
y = zeros(n+1,m);
y(1,1:m) = yo(1:m);
dydx = zeros(1,m);
ytemp = zeros(1,m);
k1 = zeros(1,m);
k2 = zeros(1,m);
k3 = zeros(1,m);
k4 = zeros(1,m);
for i =   1:1:n
   x1 = x(i);
   ytemp(1:m) = y(i,1:m);
   k1(1:m) = funkeval(x1,ytemp);
   x2 = x(i) + 0.5*dx;
   ytemp(1:m) = y(i,1:m) + 0.5*dx*k1(1:m);
   k2(1:m) = funkeval(x2,ytemp);
   x3 = x(i) + 0.5*dx;
   ytemp(1:m) = y(i,1:m) + 0.5*dx*k2(1:m);
   k3(1:m) = funkeval(x3,ytemp);
   x4 = x(i) + dx;
   ytemp(1:m) = y(i,1:m) + dx*k3(1:m);
   k4(1:m) = funkeval(x4,ytemp);
   dydx(1:m) = 1.0/6.0*(k1(1:m) + 2.0*k2(1:m) + 2.0*k3(1:m) + k4(1:m));
   y(i+1,1:m) = y(i,1:m) + dx*dydx(1:m);
end
close all;
iplot = 1;
if (iplot == 1)
   for i = 1:1:m
```

Ordinary Differential Equations - 129

```
      color_index = get_plot_color(i);
      plot (x(:),y(:,i),color_index);
      hold on;
   end
   hold off;
   xlabel( 'x' );
   ylabel ( 'y' );
   legend (int2str([1:m]'));
end
fid = fopen('rk4n_out.txt','w');
fprintf(fid,'x    y(1) ... y(m) \n');
for i = 1:1:n+1
   fprintf(fid,'%23.15e ', x(i));
   for j = 1:1:m
      fprintf(fid,'%23.15e ', y(i,j));
   end
   fprintf(fid,' \n');
end
fclose(fid);

function dydx = funkeval(x,y);
dydx(1) = -1.0*y(1) - 2.0*y(2) - 0.5*y(3);
dydx(2) = -0.1*y(1) - 4.0*y(2) - 0.5*y(3);
dydx(3) = -0.5*y(1) - 0.4*y(2) - 0.2*y(3);
```

An example of using rk4n_short is given below.

```
» [x,y]=rk4n(100,0,10,[1,0,2]);
```

Note that the fourth input argument, yo, is now a vector. This program generates outputs in three forms. First, the x vector and y matrix are stored in memory and can be directly accessed. Second, the program generates a plot of y vs. x. Third, the program generates an output file, *eulern_out.txt*, that contains x and y in tabulated form.

Also note that this code calls an ancillary function for plotting, *get_plot_color*, which is reproduced above. The function assigns a different color to each function so that the curves are distinguishable on the graph. This function can be included at the bottom of the file rk4n_short.m.

Code 7.5. Newton-Raphson/Runge-Kutta Method – 2 ODEs BVP (nrnd1 & rk4n)

No new codes were used to solve the BVP problem. Instead the input files for the Newton-Raphson with numerical derivatives (nrnd1.m) and the classical fourth-order Runge-Kutta method for a system of ODEs (rk4n.m) were modified.

In the file rk4n.m, we entered the ODEs

```
function dydx = funkeval(x,y);
dydx(1) = y(2);
dydx(2) = -1.0*y(1) - 0.5*y(2);
```

In the file nrnd1.m, the input function was specified as

```
function f = funkeval(x)
xo = 0;
yo_1 = 1;
yo_2 = x;
xf = 2.0;
yf = 1.0;
n = 100;
[x,y]=rk4n(n,xo,xf,[yo_1,yo_2]);
yf_calc = y(n+1,1);
f = yf_calc-yf;
```

In this input file, we call the Runge-Kutta routine. We provide the given boundary conditions. We allow the initial condition for the second function to vary with each iteration, since it is the unknown in the Newton-Raphson procedure.

An example of using solving this BVP is given below.

```
» [x0,err] = nrnd1(0.1)
icount = 1 xold = 1.000000e-001 f = -1.012145e+000 df = 5.850002e-001 xnew = 1.830161e+000  err = 1.000000e+002

icount = 2 xold = 1.830161e+000 f = -2.167155e-013 df = 5.850002e-001 xnew = 1.830161e+000  err = 2.023704e-013

x0 =     1.8302
err =    2.0237e-013
```

The problem was solved in one iteration because the ODEs were linear. It took a second iteration to identify convergence. The initial value of the derivative required to solve the boundary condition is 1.8302. The rk4n.m program generates a plot of y vs. x and generates an output file, *rk4n_out.txt*, that contains x and y in tabulated form. However, be warned that this plot and table of data correspond to the last call of the Runge-Kutta routine by the Newton-Raphson code, which would have been to evaluate the numerical derivative. Therefore, this plot does not correspond exactly to the solution. To generate the solution, run the Runge-Kutta code with the appropriate initial condition.

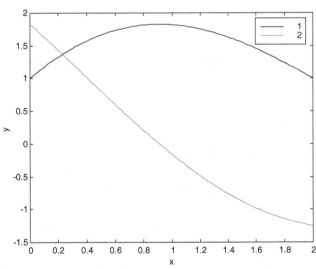

Figure 7.3. Solution to boundary value problem. $\frac{d^2 y(x)}{dx^2} = -y(x) - \frac{1}{2}\frac{dy(x)}{dx}$ subject to the boundary conditions $y(0)=1$ and $y(2)=1$. The two curves correspond to y (black) and the first derivative (red).

```
» [x,y]=rk4n(100,0,2,[1,1.8302]);
```

The resulting plot is shown in Figure 7.3. It demonstrates that the final value of y is indeed the specified value of 1.

7.9. Problems

Problem 7.1.

Consider the initial value problem:

$$\frac{dy}{dx} + a(x)y = b(x)$$

where we have an initial condition of the form:

$$y(x = x_o) = y_o$$

with the specific values given by:

$$a(x) = 2, \quad b(x) = x\sin(3x), \quad y(x = 0) = 1$$

(a) Analytically solve for y(x) from x = 0 to 4.
(b) Plot the analytical solution.
(c) Use Euler with a time step of 0.4
(d) Use Euler with a time step of 0.04
(e) Use Runge-Kutta with a time step of 0.4
(f) Use Runge-Kutta with a time step of 0.04
(g) Compare the relative error of the Euler estimate of y(x=4) for both sized steps. Explain.
(h) Compare the relative error of the Runge-Kutta estimate of y(x=4) for both sized steps. Explain.
(i) Compare the relative errors of the Euler and Runge-Kutta estimates of y(x=4) for a time step of 0.04. Explain.
(j) Compare the relative errors of the Runge-Kutta estimates of y(x=2) and y(x=4) for a time step of 0.04. Explain.

Problem 7.2.

Consider the system of non-linear ordinary differential equation:

$$\frac{dy_1}{dx} = \frac{k_{11} y_1}{k_{12} y_2} + b_1 x$$

$$\frac{dy_2}{dx} = \sqrt{k_{21} y_1 + k_{22} y_2 + b_2 x}$$

with the initial conditions

$$y_1(x=0) = y_{1o} = 1.0$$
$$y_2(x=0) = y_{2o} = 0.1$$

where

$$k = \begin{bmatrix} -0.1 & 0.1 \\ 0.03 & 0.02 \end{bmatrix} \text{ and } b = \begin{bmatrix} -0.01 \\ 0.0 \end{bmatrix}$$

(a) Determine the behavior of $y_1(x)$ and $y_2(x)$ from $0 \leq x \leq 20$.
(b) Determine the values of $y_1(x)$ and $y_2(x)$ at $x = 10$.

Problem 7.3.

The transient behavior of a continuously stirred tank reactor (CSTR) operated in adiabatic mode can be described by two ordinary differential equations describing the conservation of mass and energy. The independent variable is time, t. The dependent variables are the extent of reaction, x, and the temperature, T. The code provided below provides a set of parameters necessary for a complete, sample case study. Model the transient behavior of this reactor from a variety of initial states, including

(a) x = 0.01 & T = 300 K
(b) x = 0.99 & T = 300 K
(c) x = 0.01 & T = 400 K
(d) x = 0.99 & T = 400 K

```
x = y(1);            % extent of reaction
T = y(2);            % Temperature K
Cin = 3.0;           % inlet concentration mol/l
C = Cin*(1-x);       % concentration
Q = 60e-3;           % volumetric flowrate l/s
R = 8.314;           % gas constant J/mol/K
Ea = 62800;          % activation energy J/mol
```

```
ko = 4.48e+6;          % reaction rate prefactor 1/s
k = ko*exp(-Ea/(R*T)); % reaction rate constant 1/s
V = 10;                % reactor volume l
Cp = 4.19e3;           % heat capacity J/kg/K
Tin = 298;             % inlet feed temperature K
Tref = 298;            % thermodynamic reference temperature K
DHr = -2.09e5;         % heat of rxn J/mol
rho = 1.0;             % density kg/l
dydt(1) = 1/V*(Q*Cin - Q*C - k*C*V);   % mass balance mol/s
dydt(2) = 1/(Cp*rho*V)*(Q*Cp*rho*Tin  - Q*Cp*rho*T - DHr*k*C*V);       % NRG balance J/s
dydt(1) = -1/Cin*dydt(1); % convert conc. to extent
```

Chapter 8. Optimization

8.1. Introduction

The contents of the first seven chapters prepare the scientist and engineer to accomplish an extraordinarily broad set of tasks—solving systems of algebraic equations, either linear or nonlinear, solving systems of ordinary differential equations, numerical differentiation, integration and linear regression. For many undergraduate introductions to numerical methods, this would be sufficient. However there is one task that undergraduates frequently face that we have not yet covered and that task is optimization. Optimization means finding a maximum or minimum. In mathematical terms, optimization means finding where the derivative is zero.

$$\frac{df(x)}{dx} = 0 \tag{8.1}$$

Typically, the function, $f(x)$, is called the objective function, since the objective is to optimize it.

It is true that linear regression is one kind of optimization. We observed in the derivation of the regression techniques in Chapter 2 that we analytically differentiated the sum of the squares of the error (SSE) with respect to the regression parameters and then set the expressions for those partial derivatives to zero, resulting in a system of linear algebraic equations. If we can linearize the model (that is, massage the equation so that the unknown parameters appear in a linear form), then this is absolutely the way to proceed because it is always much easier to solve a system of linear algebraic equations than a system of nonlinear algebraic equations.

However, life is nonlinear and sometimes we are asked to optimize parameters for nonlinear models. To this end we provide a brief chapter on optimization of nonlinear systems. There is a vast and ever-expanding literature on optimization techniques. This quaint chapter is intended to provide a couple rudimentary techniques and introduce the student to the subject. One virtually universally acknowledged source on optimization is "Numerical Recipes" by Press et al. Students who find this chapter does not sate their curiosity are encouraged to seek out Chapter 10 of "Numerical Recipes".

8.2. Optimization vs Root-finding in One-Dimension

One can compare the goal of optimization in equation (8.1) with the goal of root-finding in equation (4.1)

$$f(x) = 0 \tag{4.1}$$

The two equations are essentially the same, both setting a function equal to zero. This motivates the idea that we can use our existing single-equation root-finding tools to optimize nonlinear equations. If the function is simple, we can perform the differentiation by hand and obtain the functional form of the derivative. At that point we can apply any single-equation root-finding technique, such as the bisection method or the Newton-Raphson method, without modification, using as an input $f'(x)$ rather than $f(x)$.

If we either cannot or will not differentiate the function analytically, we can still use the framework of the bisection method or the Newton-Raphson method where we use the finite difference formula to provide the first derivative of the function. If we are using a technique like the bisection method, that is all we require as input. If we are using a technique like the Newton-Raphson method, which requires derivatives of the function, then we shall require the second derivative as well. Fortunately, in Chapter 3, we provided finite difference formulae for the second derivative as well.

Later in this chapter, we provide subroutines for using the bisection and Newton-Raphson method for one-dimensional optimization. The only change in the bisection code necessary to convert it from a root-finding routine to an optimization routine is that, where we previously evaluated the function at the brackets, we now evaluate the first derivative of the function at the brackets using a finite difference formula. The only changes in the Newton-Raphson with Numerical derivatives method to convert it from a root-finding routine to an optimization routine are that, (i) where we previously evaluated the function, we now evaluate the first derivative of the function using a finite difference formula and (ii) where we previously evaluated the first derivative of the function, we now evaluate the second derivative of the function using a finite difference formula.

Example 8.1. Bisection Method

Consider the single nonlinear algebraic equation as our objective function,

$$f(x) = \exp(x) - sqrt(x) \tag{8.2}$$

Although the derivative of this function can easily be evaluated, we will not do so for the sake of this application. We will take as our brackets,

$$x_- = 0.1 \text{ and } x_+ = 0.55.$$

How did we find these brackets? It was either by trial and error or we plotted $f(x)$ vs x to obtain some idea where the minimum was. For optimization, we need the **slope** to be negative at x_- and the slope to be positive at x_+. We will use a relative error on x as the criterion for convergence and we will set our tolerance at 10^{-6}.

	x_-	x_+	$f'(x_-)$	$f'(x_+)$	error
1	0.100000	0.550000	-4.76E-01	1.06E+00	8.18E-01
2	0.100000	0.325000	-4.76E-01	5.07E-01	6.92E-01
3	0.100000	0.212500	-4.76E-01	1.52E-01	5.29E-01
4	0.156250	0.212500	-9.58E-02	1.52E-01	2.65E-01
5	0.156250	0.184375	-9.58E-02	3.80E-02	1.53E-01
6	0.170313	0.184375	-2.59E-02	3.80E-02	7.63E-02
7	0.170313	0.177344	-2.59E-02	6.72E-03	3.96E-02
8	0.173828	0.177344	-9.41E-03	6.72E-03	1.98E-02
9	0.175586	0.177344	-1.30E-03	6.72E-03	9.91E-03
10	0.175586	0.176465	-1.30E-03	2.72E-03	4.98E-03
11	0.175586	0.176025	-1.30E-03	7.12E-04	2.50E-03
12	0.175806	0.176025	-2.95E-04	7.12E-04	1.25E-03
13	0.175806	0.175916	-2.95E-04	2.09E-04	6.25E-04
14	0.175861	0.175916	-4.30E-05	2.09E-04	3.12E-04
15	0.175861	0.175888	-4.30E-05	8.29E-05	1.56E-04
16	0.175861	0.175874	-4.30E-05	1.99E-05	7.81E-05
17	0.175868	0.175874	-1.15E-05	1.99E-05	3.90E-05
18	0.175868	0.175871	-1.15E-05	4.22E-06	1.95E-05
19	0.175869	0.175871	-3.65E-06	4.22E-06	9.76E-06
20	0.175869	0.175870	-3.65E-06	2.88E-07	4.88E-06
21	0.175870	0.175870	-1.68E-06	2.88E-07	2.44E-06

So in 21 iterations, we see the optimum value lies at $x = 0.175870$. The bisection method guarantees a root. However, since we are invoking a finite difference formula to estimate the derivative of the function, we are subject to the limitations in the accuracy of the numerical differentiation. As for virtually any numerical method, it is conceivable that ill-posed problems exist for this procedure will not converge.

Example 8.2. Newton-Raphson with Numerical Derivatives Method

Consider the same objective function as used in the previous example. We will provide an initial guess for the Newton-Raphson method of $x = 1$. We will use a relative error on x as the criterion for convergence and we will set our tolerance at 10^{-6}.

	x_{old}	$f(x_{old})$	$f'(x_{old})$	x_{new}	error
1	2.000000	7.035625	7.48E+00	1.06E+00	1.00E+02
2	1.059095	2.397952	3.11E+00	2.89E-01	2.67E+00
3	0.288832	0.404506	2.95E+00	1.52E-01	9.06E-01
4	0.151500	-0.121024	5.40E+00	1.74E-01	1.29E-01
5	0.173898	-0.009088	4.64E+00	1.76E-01	1.11E-02
6	0.175858	-0.000054	4.58E+00	1.76E-01	6.75E-05
7	0.175870	0.000000	4.58E+00	1.76E-01	3.25E-09

So we converged to 0.175870 in only seven iterations.

8.4. Other One Dimensional Optimization Techniques

Nonlinear optimization can be a tricky business. There is always the possibility of falling into local minimum rather than the desired global minimum. Therefore, an exhaustive optimization search involves numerous initial guesses. If the various initial guesses lead to the different minima, then the global minimum is the one with the lowest value of the objective function.

Sometimes it is also worth trying different optimization techniques in order to avoid pitfalls of one method. For this reason, two additional optimization techniques are described below. The practical procedure in optimization is thus to try a given method. If it fails to converge, then we move on to another method. We naturally try the fastest methods (with quadratic convergence) first. If that fails, we then try the slower methods. Once we have a minimum, it is often useful to confirm the minimum by starting the optimization again (with the same method and with different methods) at a point very near the minimum to confirm the existence of the minimum (as opposed to simply having run out of iterations before convergence). Thus we might try to optimize using the Newton-Raphson method first. If it fails, we might then move on to Brent's method, described below.

A seminal resource in Numerical Methods is the book "Numerical Recipes" by Press, Teukolsky, Vetterling & Flannery. They provide various excellent routines for one-dimensional optimization. The codes are provided in either Fortran or C, depending on the version of the text book. Because these tools are so refined, we have translated some of the routines into Matlab and made the translated codes available on the course website. None of the codes translated from "Numerical Recipes" are reproduced in this book.

Relevant to the topic of one-dimensional optimization, the Brent's method of one-dimensional optimization (brent.m) and Brent's method of one-dimensional optimization with numerical derivatives (dbrent.m) are provided in Matlab. These methods incorporate not simply theory but a variety of rules of thumb and tricks of the trade for optimization, gained over many years of experience by the authors of "Numerical Recipes". The two codes, brent.m and dbrent.m require three brackets. Thus the translation of the bracket generating routine for one-dimensional optimization, mnbrak1.m is also included on the course website. These routines require that the objective function be entered in a file funkeval.m.

Example 8.3. Brent Method

Consider the same objective function in equation (8.2). We first create the objective function, funkeval.m,

```
function f = funkeval(x);
f = exp(x) - sqrt(x);
```

We next generate brackets,

```
» [ax, bx, cx] = mnbrak1(0.1,0.2,'min')

ax =     0.1000
bx =     0.2000
```

```
cx =     0.3618
```

Note that the first two input arguments in mnbrak1 are two values of x, hopefully close to the root. The third argument, 'min', specifies minimization. The outputs are three brackets required by both brent.m and dbrent.m.

We next run the optimization code. For example, if we use brent.m, with a relative tolerance on x (fourth argument) of 10^{-6} and the print option (fifth argument) turned on (1 = on, 0 = off), we have

```
» [f,xmin] = brent(ax,bx,cx,1.0e-6,1,'min');
boundary 1:    a = 1.76e-001 f(a) =   7.73e-001
boundary 2:    b = 1.76e-001 f(b) =   7.73e-001
best guess:    x = 1.76e-001 f(x) =   7.73e-001
2nd best guess:  w = 1.76e-001 f(w) =   7.73e-001
previous w:    v = 1.76e-001 f(v) =   7.73e-001
most recent point:  u = 1.76e-001 f(u) =   7.73e-001
error = 4.35e-008 iter =     28

ANSWER = 1.758669e-001
```

So in 28 iterations, Brent's method found the minimum at 0.17587 with an error of 4.35×10^{-8}.

Alternatively, we could use the dbrent.m code.

```
» [f,xmin] = dbrent(ax,bx,cx,1.0e-6,1,'min');
boundary 1:    a = 1.76e-001 f(a) =   7.73e-001
boundary 2:    b = 1.76e-001 f(b) =   7.73e-001
best guess:    x = 1.76e-001 f(x) =   7.73e-001
2nd best guess:  w = 1.76e-001 f(w) =   7.73e-001
previous w:    v = 1.76e-001 f(v) =   7.73e-001
most recent point:  u = 1.76e-001 f(u) =   7.73e-001
error = 4.05e-008 iter =     20

ANSWER = 1.758669e-001
```

So in 20 iterations, Brent's method with derivatives found the same minimum at 0.17587 with an error of 4.05×10^{-8}.

8.5. Multivariate Nonlinear Optimization

Real systems and materials are not only nonlinear but also are dependent on multiple parameters. Therefore multivariate nonlinear optimization is a task that is practically necessary. The objective in multivariate nonlinear optimization is to minimize one objective function with respect to many variables,

$$\left(\frac{\partial f(x_1, x_2 \ldots x_n)}{\partial x_i}\right)_{x_{j \neq i}} = 0 \qquad \text{for } i = 1 \text{ to } n \tag{8.3}$$

It is often said that multivariate nonlinear optimization is an art not a science. The challenge of finding any minimum in multi-dimensional space can be difficult. The challenge of routinely finding a global minimum in multi-dimensional space is exceedingly difficult. This task is still an active area of research.

In this chapter we will provide discussion of three useful techniques for multivariate nonlinear optimization. The first is the simple conversion of a root-finding technique. The latter two are optimization techniques with codes translated from "Numerical Recipes".

8.6. Optimization vs Root-finding in Multiple Dimensions

Just as we converted the single-variable version of the Newton-Raphson with numerical derivatives method from a root-finding technique to an optimization technique, so too can we convert the multivariate Newton-Raphson method with numerical derivatives method from a root-finding technique to an optimization technique.

In this case, we have one objective function, $f_{obj}(\underline{x})$, which a function of n variables. At the kth iteration, the ith element of the residual in the Newton-Raphson method is first partial derivative of the objective function with respect to variable x_i, evaluated at the current values of $\underline{x}^{(k)}$,

$$R_i^{(k)} = \left(\frac{\partial f_{obj}}{\partial x_i}\right)_{x_{m \neq i}} \Bigg|_{\underline{x}^{(k)}} \tag{8.4}$$

The i,j element of the Jacobian matrix at the kth iteration is the matrix of second partial derivatives, which in mathematical is called the Hessian matrix,

$$J_{i,j}^{(k)} = \left(\frac{\partial}{\partial x_j}\left(\frac{\partial f_{obj}}{\partial x_i}\right)_{x_{m \neq i}}\right)_{x_{m \neq j}} \Bigg|_{\underline{x}^{(k)}} \tag{8.5}$$

By inspection, the Hessian matrix is symmetric.

It is likely that the derivatives required in the residual vector and Jacobian matrix will be evaluated numerically. To do so, we employ the centered-finite difference formulae for first and second partial derivatives provided in Section 3.4.

Example 8.4. Multivariate Newton-Raphson with Numerical Derivatives Method
Consider the objective function,

$$f_{obj}(x_1, x_2) = (x_1 - 3)^2 + (x_2 - 7)^2 + x_1 x_2^2 - x_1^2 x_2 + 4 \tag{8.6}$$

If we provide an initial guess of $(x_1, x_2) = (4, 8)$, we find that this method produces the following results

iteration	x_1	x_2	error on residual	error on x
0	4	8	-	-
1	2.1373	4.4902	35.4	2.81
2	1.3797	3.5274	6.81	0.866
3	1.2308	3.4759	0.73	0.111
4	1.2302	3.4781	1.02e-02	1.53e-03
5	1.2302	3.4781	5.56e-6	8.22e-07

So in 5 iterations, the Newton-Raphson method with numerical derivatives found the same minimum at $x_1 = 1.23$ and $x_2 = 3.47$ with a relative error on x of 8.22×10^{-7} and an RMS error on the residual of 5.56×10^{-6}. The value of the objective function at the minimum is 29.154, determined by substitution of the converged solution into the objective function.

8.7. Other Multivariate Optimization Techniques

As was the case for the single-variable nonlinear optimization, a seminal resource in multivariate nonlinear optimization is the book "Numerical Recipes" by Press, Teukolsky, Vetterling & Flannery. They provide various excellent routines for multi-dimensional optimization. The codes are provided in either Fortran or C, depending on the version of the text book. Because these tools are so refined, we have translated some of the routines into Matlab and made the translated codes available on the course website. None of the codes translated from "Numerical Recipes" are reproduced in this book.

We discuss two completely different approaches to multivariate nonlinear optimization presented in "Numerical Recipes". The first is the called either the "Nelder and Mead's Downhill Simplex Method" or the "amoeba method". In the application of the amoeba method to an n-dimensional optimization problem, one creates an n-dimensional volume using n+1 points. For example a two-dimensional volume (typically called an area) can be created from 3 points. The resulting two-dimensional shape is called a triangle. Similarly, a three-dimensional volume can be created from 4 points. The resulting three-dimensional shape is called a tetrahedron. The same concept applies to n-dimensional space. One creates this n-dimensional object and then allows it to explore n-dimensional space through a series of reflection, contraction and extrapolation operations. The process can be liked to an amoeba who moves away from its least favorable of the n+1 points by extending a pseudopod in the opposite direction (reflection). If it likes what it finds (a more hospitable environment, in this case indicated by a lower value of the objective function), it moves even further (extrapolation). If it doesn't like what it finds, it shrinks away (contraction).

Optimization - 141

Eventually all n+1 points are within a given tolerance of the minimum and the process is converged. The advantage of this method is that it is simple and hard to crash. The disadvantage is that it is slow and can require thousands or hundreds of thousands of iterations to converge. The code amoeba.m that appears on the course website is a translation of the Fortran version of the code that appears in "Numerical Recipes". Because the amoeba method is slow, it should be used only as a last resort, when other quicker methods have failed to find the optimum. This routine require that the objective function be entered in a file funkeval.m.

Example 8.5. Amoeba Method
Consider the objective function,

$$f_{obj}(x_1, x_2) = (x_1 - 3)^2 + (x_2 - 7)^2 + x_1^2 x_2^2 + 4 \qquad (8.7)$$

For this two-dimensional problem, the amoeba method requires three initial guesses. We only want to provide one initial guess. Therefore, the code in general generates an additional n initial guess to the one guess provided by increase each variable in term by some fraction. For example, if we provide an initial guess of $(x_1, x_2) = (1,3)$ and our factor is 1%, then our additional two initial guesses are $(x_1, x_2) = (1.01, 3)$ and $(x_1, x_2) = (1, 3.03)$. The initial values of the objective function for these three initial guesses are respectively, 33, 33.141 and 32.942. We set two tolerances in the amoeba code, one for the unknowns and one for the objective function, both are set to 10^{-6} in this example. These values constrain the range of x and f_{obj} values across the n+1 points.
The code is issued at the MATLAB command line prompt with the following command:

```
» [f,x] = amoeba([1; 3],1.0e-6,1.0e-6)
```

In the following table values of x and f_{obj} that correspond to the best of the n+1 points are provided by iteration. The errors on x and f_{obj} are also reported.

iteration	x_1	x_2	f_{obj}	error on x	error on f_{obj}
1	1.00000000	3.03000000	32.941800	9.95E-03	6.03E-03
2	0.98000000	3.04500000	32.627278	1.77E-02	1.14E-02
3	0.97000000	3.11250000	32.348672	2.87E-02	1.82E-02
4	0.92500000	3.17625000	31.558717	5.06E-02	3.33E-02
5	0.88250000	3.34312500	30.560857	8.38E-02	5.68E-02
...	...	...	...	...	...
62	0.06043011	6.97453100	12.819358	5.27E-06	5.58E-13
63	0.06043011	6.97453100	12.819358	2.88E-06	5.31E-13
64	0.06043011	6.97453100	12.819358	3.36E-06	1.58E-13
65	0.06043026	6.97453030	12.819358	1.33E-06	1.21E-13
66	0.06043017	6.97452980	12.819358	6.98E-07	2.06E-14

So in 66 iterations, the amoeba method found a minimum at $x_1 = 0.6043$ and $x_2 = 6.9745$ with a relative error on x of 6.98×10^{-7} and an RMS error on the objective function of 2.06×10^{-14}. The value of the objective function at the minimum is 12.819.

The second method for multivariate nonlinear optimization that has been translated from "Numerical Recipes" is the conjugate-gradient method. The procedure of a conjugate gradient method is to perform a series of one-dimensional line minimizations. We have already been introduced to good tools for one-dimensional line minimizations earlier in the chapter. However, the directions (or gradients) of these lines do not correspond to the variable axes. Instead these gradients are selected to be as independent of (or orthogonal to) each other as possible. In the best case, the conjugate gradient method turns an n-dimensional optimization problem into a series of n one-dimensional optimization problems. Because the conjugate gradient method uses information about derivatives (obtained numerically in this code as in the Newton-Raphson code), it will converge rapidly when one is near the optimum.

Example 8.6. Conjugate Gradient Method

Consider the same objective function as was used in the previous example, equation (8.7). We again provide an initial guess of $(x_1, x_2) = (1,3)$. We set a tolerance for the relative error on x to 10^{-6} in this example. The code is issued at the MATLAB command line prompt with the following command:

```
» [f,x] = conjgrad([1;3],1.0e-6,1,'min')
```

The third and fourth arguments are a printing variable (1 for printing intermediate information from each iteration) and a key set to minimization. In the following table values of x and f_{obj} are provided by iteration. The error on x is also reported.

iteration	x_1	x_2	f_{obj}	error on x
0	1.000000	2.000000	37.000000	1.00E+02
1	-0.127925	3.691887	24.950576	6.24E+00
2	-0.220721	6.266219	16.824407	4.16E-01
3	0.060911	6.967095	12.819419	3.27E+00
4	0.060641	6.974728	12.819360	3.24E-03
5	0.060430	6.974531	12.819358	2.47E-03
6	0.060430	6.974530	12.819358	4.16E-07

So in 6 iterations, the conjugate gradient method found a minimum at $x_1 = 0.6043$ and $x_2 = 6.9745$ with a relative error on x of 4.16×10^{-7} and the value of the objective function at the minimum is 12.819.

8.8. Subroutine Codes

In this section, we provide routines for implementing the various optimization methods described above that are not translated from "Numerical Recipes". Note that these codes correspond to the theory and notation exactly as laid out in this book. These codes do not contain

Optimization - 143

extensive error checking, which would complicate the coding and defeat their purpose as learning tools. That said, these codes work and can be used to solve problems.

As before, on the course website, two entirely equivalent versions of this code are provided and are titled *code.m* and *code_short.m*. The short version is presented here. The longer version, containing instructions and serving more as a learning tool, is not presented here. The numerical mechanics of the two versions of the code are identical.

Code 8.1. Bisection Method for optimization – 1 variable (bisect_opt1_short)

```
function [x0,err] = bisect_opt1(xn,xp);
maxit = 100;
tol = 1.0e-6;
err = 100.0;
icount = 0;
h = min(0.01*xn,0.01);
fn = dfunkeval(xn,h);
h = min(0.01*xp,0.01);
fp = dfunkeval(xp,h);
while (err > tol & icount <= maxit)
   icount = icount + 1;
   xmid = (xn + xp)/2;
   h = min(0.01*xp,0.01);
   fmid = dfunkeval(xmid,h);
   if (fmid > 0)
      fp = fmid;
      xp = xmid;
   else
      fn = fmid;
      xn = xmid;
   end
   err = abs((xp - xn)/xp);
   fprintf(1,'i = %i xn = %e xp = %e fn = %e fp = %e err = %e \n',icount, xn, xp, fn, fp, err);
end
x0 = xmid;
if (icount >= maxit)
   fprintf(1,'Sorry.  You did not converge in %i iterations.\n',maxit);
   fprintf(1,'The final value of x was %e \n', x0);
end

function df = dfunkeval(x,h)
fp = funkeval(x+h);
fn = funkeval(x-h);
df = (fp - fn)/(2*h);

function f = funkeval(x)
f = exp(x)-sqrt(x);
```

An example of using bisect_opt1_short is given below.

```
» [x0,err] = bisect_opt1_short(0.1,1);

i = 1 xn = 1.000000e-001 xp = 5.500000e-001 fn = -4.759875e-001 fp =
1.059054e+000 err = 8.181818e-001
...
i = 23 xn = 1.758699e-001 xp = 1.758700e-001 fn = -2.037045e-007 fp =
2.877649e-007 err = 6.100434e-007

x0 =    0.17586992979050
err =    6.100434297642754e-007
```

So in 23 iterations, the bisection method found the same minimum at 0.17587 with an error of 6.10×10^{-7}.

Code 8.2. Newton-Raphson Method with numerical derivatives for optimization – 1 variable (nrnd_opt1_short)

```
function [x0,err] = nrnd_opt1(x0);
maxit = 100;
tol = 1.0e-6;
err = 100.0;
icount = 0;
xold =x0;
while (err > tol & icount <= maxit)
   icount = icount + 1;
   h = min(0.01*xold,0.01);
   f = dfunkeval(xold,h);
   df = d2funkeval(xold,h);
   xnew = xold - f/df;
   if (icount > 1)
      err = abs((xnew - xold)/xnew);
   end
   fprintf(1,'icount = %i xold = %e f = %e df = %e xnew = %e  err = %e \n',icount, xold, f, df, xnew, err);
   xold = xnew;
end
x0 = xnew;
if (icount >= maxit)
   % you ran out of iterations
   fprintf(1,'Sorry.  You did not converge in %i iterations.\n',maxit);
   fprintf(1,'The final value of x was %e \n', x0);
end

function df = dfunkeval(x,h)
fp = funkeval(x+h);
fn = funkeval(x-h);
df = (fp - fn)/(2*h);

function d2f = d2funkeval(x,h)
fp = funkeval(x+h);
```

Optimization - 145

```
fo = funkeval(x);
fn = funkeval(x-h);
d2f = (fp - 2*fo + fn)/(h*h);

function f = funkeval(x)
f = exp(x)-sqrt(x);
```

An example of using nrnd_opt1_short is given below.

```
» [x0,err] = nrnd_opt1_short(2)
icount = 1 xold = 2.000000e+000 f = 7.035625e+000 df = 7.477507e+000 xnew = 1.059095e+000   err = 1.000000e+002
...
icount = 7 xold = 1.758700e-001 f = -2.621053e-009 df = 4.582021e+000 xnew = 1.758700e-001   err = 3.252574e-009

x0 =    0.17586997424458
err =   3.252573753217557e-009
```

So in 7 iterations, the Newton-Raphson method with numerical derivatives found the same minimum at 0.17587 with an error of 3.25×10^{-9}.

Code 8.3. Newton-Raphson Method with numerical derivatives for optimization – n variables (nrnd_optn_short)

```
function [x,err,f] = nrnd_optn(x0,tol,iprint)
maxit = 1000;
n = max(size(x0));
Residual = zeros(n,1);
Jacobian = zeros(n,n);
InvJ = zeros(n,n);
dx = zeros(n,1);
x = zeros(n,1);
xold = zeros(n,1);
dxcon = zeros(n,1);
dxcon(1:n) = 0.01;
x = x0;
err = 100.0;
iter = 0;
while ( err > tol )
   for j = 1:1:n
      dx(j) = min(dxcon(j)*x(j),dxcon(j));
   end
   Residual = dfunkeval(x,dx,n);
   Jacobian = d2funkeval(x,dx,n);
   xold = x;
   invJ = inv(Jacobian);
   deltax = -invJ*Residual;
   for j = 1:1:n
```

```
            x(j) = xold(j) + deltax(j);
        end
        iter = iter +1;
        err = sqrt( sum(deltax.^2) /n  ) ;
        f = sqrt(sum(Residual.*Residual)/n);
        if (iprint == 1)
            fprintf (1,'iter = %4i, err = %9.2e f = %9.2e \n ', iter, err, f);
        end
        if ( iter > maxit)
            Residual
            error ('maximum number of iterations exceeded');
        end
end

function df = dfunkeval(x,dx,n)
df = zeros(n,1);
for i = 1:1:n
    xtemp(1:n) = x(1:n);
    xtemp(i) = x(i) + dx(i);
    fp = funkeval(xtemp);
    xtemp(i) = x(i) - dx(i);
    fn = funkeval(xtemp);
    df(i) = (fp - fn)/(2*dx(i));
end

function Jacobian = d2funkeval(x,dx,n)
Jacobian = zeros(n,n);
for i = 1:1:n
    xtemp(1:n) = x(1:n);
    xtemp(i) = x(i) + dx(i);
    fp = funkeval(xtemp);
    xtemp(i) = x(i);
    fo = funkeval(xtemp);
    xtemp(i) = x(i) - dx(i);
    fn = funkeval(xtemp);
    Jacobian(i,i) = (fp - 2*fo + fn)/(dx(i)*dx(i));
end
for i = 1:1:n-1
    for j = i+1:1:n
        xtemp(1:n) = x(1:n);
        xtemp(i) = x(i) + dx(i);
        xtemp(j) = x(j) + dx(i);
        fpp = funkeval(xtemp);
        xtemp(i) = x(i) + dx(i);
        xtemp(j) = x(j) - dx(i);
        fpn = funkeval(xtemp);
        xtemp(i) = x(i) - dx(i);
        xtemp(j) = x(j) + dx(i);
        fnp = funkeval(xtemp);
        xtemp(i) = x(i) - dx(i);
        xtemp(j) = x(j) - dx(i);
        fnn = funkeval(xtemp);
        Jacobian(i,j) = (fpp - fpn -fnp + fnn)/(4*dx(i)*dx(j));
```

```
    end
end
for i = 2:1:n
    for j = 1:1:i-1
        Jacobian(i,j) = Jacobian(j,i);
    end
end

function fobj = funkeval(x)
fobj = (x(1)-3)^2 + (x(2)-7)^2 + x(1)*x(2)^2 - x(1)^2*x(2) + 4;
```

An example of using nrnd_optn_short is given below.

```
» [x,err,f] = nrnd_optn_short([4,8],1.0e-6,1)
iter =    1, err = 2.81e+000  f = 3.54e+001
 iter =   2, err = 8.66e-001  f = 6.81e+000
 iter =   3, err = 1.11e-001  f = 7.30e-001
 iter =   4, err = 1.58e-003  f = 1.02e-002
 iter =   5, err = 8.22e-007  f = 5.56e-006

x =   1.23017202549369   3.47805528788098
err =   8.220328680806535e-007
f =    5.555092652529348e-006
```

So in 5 iterations, the Newton-Raphson method with numerical derivatives found the same minimum at $x_1 = 1.23$ and $x_2 = 3.47$ with a relative error on x of 8.22×10^{-7} and an RMS error on the residual of 5.56×10^{-6}. The value of the objective function at the minimum is 29.154, determined by substitution of the converged solution into the objective function.

8.9. Problems

Problem 8.1.

Consider the rate equation

$$rate = k_o e^{-\frac{E_a}{RT+C}} \quad [\text{mol/s}]$$

where

prefactor: k_o [mol/sec]

activation energy for reaction: E_a [Joules/mole]

non-Arrhenius parameter: C [Joules/mole]

constant: $R = 8.314$ [Joules/mole/K]

temperature: T [K]

Determine the rate constants, E_a, k_o and C, from experimental data. The rate as a function of temperature is given in tabular form below.

For initial guesses, use the knowledge that E_a should be on the order of 10,000 J/mol, k_o should be on the order of 100,000 mol/s and C should be on the order of 1000 J/mol.

Use whatever method you prefer. If you use the amoeba method, set the initial volume space to 50% of the initial guess and set your tolerance for both x and f to 10^{-8} or less. Use the RMS (root-mean-square error) as the objective function.

$$f_{obj} = \sqrt{\frac{1}{n_{data}} \sum_{i=1}^{n_{data}} \left(rate_i^{exp} - rate_i^{mod}\right)^2}$$

temperature (K)	rate (mol/sec)
300	1.33E-02
320	2.87E-02
340	6.61E-02
360	1.24E-01
380	2.31E-01
400	3.98E-01
420	6.71E-01
440	9.91E-01
460	1.57E+00
480	2.20E+00
500	3.69E+00
520	4.43E+00
540	6.06E+00
560	8.49E+00
580	1.21E+01
600	1.76E+01

References

Chapra, S.C., Canale, R.P., <u>Numerical Methods for Engineers</u>, 2nd Ed., McGraw-Hill, New York, 1988.

Montgomery, D.C., Runger, D.C., <u>Applied Statistics and Probability for Engineers</u>, 2nd Edition, John Wiley & Sons, New York, 1999.

Press, W.H., Teukolsky, S.A., Vetterling, W.T., Flannery, B.P., <u>Numerical Recipes in Fortran 77: The Art of Scientific Computing</u>, 2nd Edition, Volume 1 of Fortran Numerical Recipes, Cambridge University Press, 1992.

Walpole, R.E., Myers, R.H, Myers, S.L., <u>Probability and Statistics for Engineers and Scientists</u>, 6th Edition, Prentice Hall, Upper Saddle River, New Jersey, 1998.

CPSIA information can be obtained
at www.ICGtesting.com
Printed in the USA
LVHW10s1528280818
588389LV00004BA/318/P